TMI 25 Years Later

TMI 25 Years Later

The Three Mile Island Nuclear Power Plant Accident and Its Impact

Bonnie A. Osif
Anthony J. Baratta
Thomas W. Conkling

The Pennsylvania State University Press
University Park, Pennsylvania

Library of Congress Cataloging-in-Publication Data

Osif, Bonnie A.
    TMI 25 years later : the Three Mile Island nuclear power plant accident and its
    impact / Bonnie A. Osif, Anthony J. Baratta, Thomas W. Conkling.
        p.   cm.
Includes bibliographical references and index.
ISBN 0-271-02383-X (hardcover : alk. paper)
1. Three Mile Island Nuclear Power Plant (Pa.)
2. Nuclear power plants—Pennsylvania—Three Mile Island—Accidents.
3. Nuclear power plants—Accidents—Social aspects—Pennsylvania—Three Mile
Island.
4. Nuclear power plants—Accidents—Health aspects—Pennsylvania—Three Mile
Island.
5. Nuclear power plants—Accidents—Economic aspects—Pennsylvania—Three
Mile Island.
I. Baratta, Anthony John, 1945–
II. Conkling, Thomas W., 1949–
III. Title.

TK1345.H37O85   2004
363.17'99'0974818—dc22
2004000490

The contents of this book reflect the views solely of the authors and are not
necessarily those of the U.S. Nuclear Regulatory Commission or of The
Pennsylvania State University.

Unless otherwise noted, all plates are from the TMI Recovery and Decontamination
Collection of the University Libraries, The Pennsylvania State University.

First paperback printing 2006

The Pennsylvania State University Press is a member of the Association of
American University Presses.

It is the policy of The Pennsylvania State University Press to use acid-free paper.
Publications on uncoated stock satisfy the minimum requirements of American
National Standard for Information Sciences-Permanence of Paper for Printed
Library Material, ANSI Z39.48-1992.

Contents

	Preface and Acknowledgments	vi
	Introduction	viii
1	Nuclear Energy Basics	1
2	The Accident	21
3	The Cleanup of TMI Unit 2	33
4	Media Coverage and Public Understanding	49
5	The Effect on the Local Community	60
6	The Impact of Three Mile Island	76
7	Energy for the Future	91
	Conclusion	114
	Appendixes	120
	Bibliography	144
	Index	150

Preface and Acknowledgments

For more than ten years, the authors of this book have maintained a Web site hosted by the Pennsylvania State University Libraries for the Three Mile Island Recovery and Decontamination Collection, which provides access to an extensive collection of reports and videotapes concerning the accident and cleanup at Three Mile Island. The collection has hundreds of reports and thousands of videos taken during the cleanup project at TMI Unit 2. The collection preserves this important and unique information.

The site was developed under grants from the former operator of TMI, General Public Utilities, and the Electric Power Research Institute, the U.S. Nuclear Regulatory Commission, and the U.S. Department of Energy. In addition to a searchable database, the site provides a short online video that summarizes the accident and a link to ask questions about the accident and its aftermath. This book is, in part, an attempt to address the range of questions and concerns that have been asked.

The book reviews the accident and its causes, the cleanup process that lasted more than a decade, and the future of energy. At all times, the writers have tried to be objective and avoid both the overdramatization of the accident and the simplification of the complexities and implications of energy choices.

A section with photographs from the collection tells the visual story of the tremendous task facing the engineers and technicians after the accident. A bibliography of pertinent works is included to provide additional information.

The chapters are organized to provide basic information on nuclear energy and reactors and to review the accident and the subsequent cleanup project based on the reports, videos, and the insights of those directly involved in the cleanup. Other chapters discuss energy options, their advantages and disadvantages, health and environmental issues, and the impact of the accident on the nuclear industry.

We hope the reader will gain insight into the significance of the events that caused the accident, the magnitude of the cleanup and the innovations to accomplish it, and the impact of the accident on the health and safety of the public and the nuclear industry. This book was written to provide the reader with a basic understanding of nuclear technology—its benefits, drawbacks, and challenges. We have tried to place the accident in perspective and provide a resource to help the general public understand an often misunderstood technology.

This book is based on the work of numerous engineers, technicians, scientists, historians, reporters, researchers, and many others. Most of these people are nameless, but they play an important role in the TMI story. The residents living near TMI who experienced the accident and its aftermath—and the hundreds of individuals who asked questions about these events—also had a significant role in this work.

We are particularly indebted to those who have helped in the preparation of this book: Samuel Walker, Ann Jensen, and Jim Ottaviani for their careful reviews of the manuscript and insightful suggestions; James J. Byrne, vice president of TMI 2, for his technical assistance; Annette D. Barnes for editorial assistance; Maureen E. Brown, Public Affairs Director, Nuclear Management Company, for her technical and editorial comments; and Pamela Leonardi and Donna E. Leonardi for their useful suggestions. We would like to acknowledge the staff of the Penn State Engineering Library for their assistance and support, Derrick Beckner for our maps, Heather Solimini for image enhancement, and Tom Minsker for his patient assistance with the formatting of the book as well as Jennifer Norton, Design and Production Manager, Penn State Press, for her assistance and Laura Reed-Morrisson for her editing of the book and ability to make difficult material understandable. A chance meeting at Mountain Home Farm with Patricia Mitchell resulted in her encouraging us to send the proposal to the Penn State Press. We would like to thank Peter Potter for shepherding us through the publication process and for his belief that this story needed to be told. Finally, we would like to thank our families for their support and understanding during the last eighteen months.

Introduction

For most people in the quiet area surrounding Middletown, Pennsylvania, March 28, 1979, dawned as an unremarkable early spring day. It was Wednesday; the weather was unexceptional. Nothing seemed out of the ordinary. It was a typical day in a pleasant rural and suburban area whose only claim to fame was that it was just outside the state capital, Harrisburg, and near the giant Hershey Chocolate Company complex.

The headline in the local paper, the *Harrisburg Patriot,* warned: "OPEC Ups Crude's Price by 9%, Allows Surcharge." Two additional articles on the front page concerned this worrisome news, while others gave some attention to the Israeli-Egyptian peace talks and several local issues.

Those headlines did not have much of an impact on most readers. Some might have thought, briefly, that rising oil prices would increase their driving costs—if not the cost of heating and lighting their homes. Perhaps the decision to build that reactor on the slim island in the Susquehanna River had been correct: nuclear power might help buffer the costs and insecurity of having so much energy dependence on those distant, somewhat mysterious nations in the Middle East. In any case, there was nothing atypical about the commute to work—many drove to the state capital ten miles away—or the bus rides to school.

Radio and television mirrored the *Patriot*'s coverage of a normal day. The media reported no earth-shattering news, nothing that was going to grab the headlines. Even those reading the big city paper, the *Philadelphia Inquirer,* saw an interesting but not dramatic headline. A front-page banner announced a major baseball event: Mets pitcher Nino Espinosa had been traded to the Philadelphia Phillies in exchange for the power hitter Richie Hebner. While of great importance to the baseball fans in Philadelphia and New York and good for discussion around the water cooler, the trade's appearance on the front page clearly indicated what a routine news day it was.

For a few people, one early-morning occurrence was not absolutely normal. Several neighbors on the river near the reactor had awakened to a loud sound at 4:00 AM. They described it as the sound of a jet engine. It was not the first time they had heard this noise, and some knew that it was the sound of steam being vented from the plant. It might be annoying, but it happened some days—a fact of life if you lived near the reactor on the island.

In the Three Mile Island (TMI) Unit 2 control room, however, there were several people who knew, as of 4:00 AM, that Wednesday would not be a typical day. And as the dark night gave way to dawn, they might have realized that it would never be viewed as a normal day again. Many Pennsylvanians, and people across the United States and around the world, would note March 28, 1979, as a pivotal day in history. Exactly how people viewed it changed rapidly as events unfolded. While history has mellowed the view of that day, its importance and far-reaching effects have not lessened.

The first announcement of something gone amiss was reported by Harrisburg radio station WKBO at 8:25 AM. The announcer mentioned a problem at the reactor but explained that there was no danger to the public. At about 9:00 AM, the Associated Press announced that the Three Mile Island nuclear reactor in Pennsylvania had had a general emergency, but the story made clear that no radiation had been released. With this short news report, the public saga of the most serious reactor incident in the United States began.

Unit 2 had developed alarming problems in the quiet morning hours. Within a short time, world attention was focused on one reactor located on an island in a little-known river in a relatively unknown area of the country—and on a technology that was poorly understood by the public.

The reactor in the Susquehanna River was one of two plants at the Three Mile Island Nuclear Generating Station. Two reactors with four cooling towers had been built on an island located in a wide, slow-moving part of the river. Unit 1 came online in September 1974 and was down for refueling that day. Unit 2 came online in December 1978, three months before the accident. Owned by Metropolitan Edison Company, Jersey Central Power and Light Company, and Pennsylvania Electric Company and operated by General Public Utilities Nuclear Corporation, the reactors provided energy for northeastern Pennsylvania and New Jersey. There had been protests against the planning and construction of the station, but most people in central

Pennsylvania had become accustomed to the four large concrete funnels reaching 370 feet into the sky as part of the everyday skyline. The construction had started eleven years earlier; the towers were visible from the state highways that paralleled the river, from many small streets, and from planes flying out of the nearby Harrisburg airport. They were part of the landscape—a part of modern life. By evening, the landscape looked much the same, but the towers and the other buildings on the narrow island represented a different perspective on modern life, one that was difficult to understand and hinted at a very real danger.

Approximately an hour after the local radio station released the story, the Associated Press reported that the reactor was leaking radiation. Reporters and citizens alike clamored for more information as the Nuclear Regulatory Commission, company representatives, and local government officials went into action.

By late afternoon, news reporters were flooding into this largely rural area. Hotels and restaurants were crowded with people who probably had never anticipated spending time in this small town outside of Harrisburg. News coverage was different in 1979 than it is today. Television was limited to the three major networks: ABC, CBS, and NBC. News networks with twenty-four-hour service and easy remote coverage were uncommon. Only a few radio stations broadcast news around the clock; the Internet, with its instant access to facts and rumors, was only a tool of a select group of scientists. The coverage was primitive by today's standards, and this only served to heighten the drama as more information (some accurate, some less so) trickled out. News conferences were called, and nuclear experts, utility spokespeople, and government officials contributed facts as well as their opinions.

The media tried to understand and interpret information that was new and rather unknown to most of them. There were conflicting stories, and the tone of the coverage varied from station to station. John Herbein, vice president of Metropolitan Edison (which owned and operated TMI), told reporters that there had been minor damage and that small traces of radiation had been released. In the late afternoon, Lieutenant Governor William Scranton told reporters that the situation was more complex than originally reported but that it represented "no danger to public health." Citizens and reporters alike were confused. How much of a crisis was there? How much of the area was in danger? Who was at risk? Where were areas of safety? How much time was there to escape should the worst happen? What *was* the worst that could happen?

Answers to these questions were not always consistent, understandable, or reassuring. Some people packed up and left. Others simply left, not taking the time to pack any of their possessions—partly to save time and partly because they were unsure whether the items had already been contaminated. A day that had started calmly ended in confusion, anger, and fear.

On March 29, as journalists continued to pour into the area, the newspaper headlines were still not overly urgent. The *New York Times* led with a story entitled "Radiation Is Released in Accident at Nuclear Plant in Pennsylvania." It was not, however, a banner headline, and the name of the plant was not mentioned until the third paragraph. The *Harrisburg Patriot* calmly stated, "Radiation Being Vented, Delay in Alert Assailed." The day after the accident, readers may have been nervous, but they were getting updates on the radio; while the engineers were struggling with the situation, they still seemed in control. Local people took the headlines seriously. While some had vacated their homes and the area around the reactor, most stayed in their homes, and those at some distance from the damaged reactor did not think they had a great deal to fear. Granted, protesters had used this accident as a chance to repeat dire warnings and predictions they had been making for years about the dangers of nuclear energy. Others pointed to a recently released film, *The China Syndrome*, to explain what might be happening. The high-profile movie starred Jane Fonda, Jack Lemmon, and Michael Douglas; it envisioned a scenario in which the core of a nuclear reactor had been uncovered and had overheated, the molten core melting through the containment building to the ground below. To the average citizen, however, what was happening on the island was a localized story, if somewhat frightening and difficult to understand.

The tone and degree of news coverage changed radically on March 30. The *New York Times* front-page headline read, "Atomic Plant Is Still Emitting Radioactivity," and the newspaper ran photographs of children playing in the foreground of a cooling tower and of chemists testing milk produced in the area. Additional stories detailed other aspects of the accident, and the lead editorial addressed "The Credibility Meltdown." The *Philadelphia Inquirer* and the *Harrisburg Patriot* followed suit, with TMI stories dominating their national coverage.

If the world had been following news accounts of Three Mile Island with calmness and patience, two announcements would change all that. The first was a miscommunication concerning a release of radioactivity. On March 30, a reading of 1200 millirem was taken above the auxiliary building after a deliberate venting. This reading, however, was reported as having occurred off site, causing fear of a serious radiation leak. The same morning, at approximately 9:50, it was announced that there was a hydrogen bubble in the reactor and that scientists feared an explosion. Suddenly, the world was faced with the unthinkable—a full-fledged nuclear disaster replete with loss of life, devastating health problems, and widespread radioactive contamination. Eyes that had glanced toward Middletown and TMI for two days were now glued to the coverage of that small island in the middle of the Susquehanna River. The nuclear age had entered a new phase, one that endures to this day.

The day of the TMI accident marks a major milestone in U.S. history, a day that shook our faith in science. Suddenly, Americans—and people around the world—realized the need to understand better the new technology and to consider the impact it would have on their everyday lives. They also had to look at their own lives, their families, and their possessions, deciding what was important and what was not.

It was a day when operators failed and when manuals did not provide procedures for such an emergency. Technology and technicians, flawed and hard-pressed, were pushed to the limit, and yet they prevailed. It was a day when the media had to consider their lack of preparation in science and technology, subjects that suddenly seemed of utmost importance in modern life. Elected officials had to face the nuclear age in their own backyards. Corporations had to answer questions—not about profit and loss, but about risk-analysis decisions.

Twenty-five years have passed since the events at Three Mile Island. Time has eroded some of the drama—especially for those who did not live through those nerve-wracking days—and provided important perspective. Many people know very little about the events that started on that March morning. Hindsight and numerous studies have provided a clear, reasoned view of what actually happened in the reactor. Today we know that there was a partial meltdown of the reactor. Fuel in the reactor overheated because of the lack of cooling water. We know that the radiation released was minor and likely caused no radiation-related health effects. We also know that the events underscored the need for highly trained personnel to operate the plants; for regulators well schooled in the technology they regulate; for a coordinated response by federal, state, and local officials; and, most of all, for an educated press and public that can understand the complex issues associated with modern technology.

This is the story of Three Mile Island and its legacy. To understand fully the incident on March 28, and the days of crisis and years of cleanup that followed, it is necessary to have a working knowledge of the basics of nuclear energy and radiation, to put energy use and production into context, and to look at the way in which these issues are covered by the media (routinely, in educating the public, as well as in times of crisis). Medical, environmental, psychological, and economic issues are important parts of the overall picture, as are other energy technologies. The facts need to be understood—and the rumors and myths put to rest. Both the unbridled promises and the unfounded fears must be put into perspective.

Nuclear energy—with its benefits and its hazards—is a major aspect of our everyday life as we turn on our lights, microwave our popcorn, and sit down in front of our television sets. Our everyday use of energy and our policy decisions concerning fuels, environment, and political alliances have roots in the events of March 28, 1979. Beyond the decisions concerning reactors are the questions that continue to plague us about health, environment, finances, safety, government, and the media's role in informing the public. The only way to answer these questions, quiet the fears, and provide for rational, considered decisions is to ensure that basic information is made available and understandable to all. Combining this information with historical perspective increases the likelihood that informed decisions can continue to be made by policymakers and the people. Our future depends upon it.

1

Nuclear Energy Basics

Nuclear energy. The term often elicits an emotional response far beyond that associated with energy from coal or hydropower. While every form of energy generation has its proponents and opponents, the stakes seem higher when the word "nuclear" is used.

Most people have only a very basic knowledge of nuclear energy, though, despite this heightened awareness. Fuel rods, radioactivity, and reactor buildings may be familiar words, but their meanings and functions are often misunderstood. The cooling tower—the tall concrete funnel clearly visible on the horizon—is a common sight. But many people confuse the cooling tower with the reactor building itself—a telling indicator of the public's general level of understanding. A number of myths further confuse the issue, thanks in part to incomplete or false information garnered from movies, television, books, and the Internet. Knowledge of the realities of nuclear energy does not guarantee that everyone will have the same opinion on the subject, but it does increase the likelihood that a debate based on accurate scientific information can take place.

Nuclear energy is just one of a number of options we have for generating electricity. It is generally considered a nonrenewable resource. Other nonrenewable resources include coal, natural gas, and oil—all substances from the earth that are limited in quantity and cannot be regenerated. Nuclear fuel differs from other resources in this group in two ways, however: it is not a fossil fuel, and it can be transmuted or changed from one element into another substance (as will be discussed below). But for all practical purposes, there is a limited amount of material suitable for use in nuclear power plants, and that material is considered nonrenewable.

There are, of course, renewable energy sources as well. These include solar, geothermal, wind, hydropower, and biomass (the latter includes wood, agricultural wastes, and other organic waste materials).

These resources, unlike those in the nonrenewable category, can be regenerated in a relatively short time. But while renewable resources have been and continue to be used as energy sources, this mainly happens on a small scale. Most of the large-scale plants that generate electricity have been based on fossil fuels.

One particular aspect of nuclear energy puts it in a class by itself. Compared to other energy sources, nuclear fuel releases immense energy. Each nuclear reaction produces about one hundred million times the energy obtained from a single chemical reaction. While the practical yield is somewhat lower than this, nuclear energy is still much more efficient than other energy resources. Indeed, a new era in electrical power generation began when the first nuclear power plant was brought online in Shippingport, Pennsylvania, in 1957.

The Atomic Basis for Nuclear Power

Nuclear power generation is based on a simple fact of physics: a great deal of energy is released when heavy atoms are split. (A heavy atom is an element with a high number of neutrons and protons, such as uranium.) This energy can be harnessed to produce electricity. The term "atom" comes from the Greek word for "indivisible." More than 2,500 years ago, Leucippus and Democritus conjectured that there was a basic unit of matter that could not be divided. This unit, they thought, made up the structure of the entire universe. While it has since been proven that the atom is not indivisible, it *is* a basic unit of matter—and the basis for the theories that led to nuclear power.

In the classical model of the atom, electrons move around the nucleus much like the planets around the sun. The nucleus is made up of neutrons and protons. Protons are positively charged, and neutrons have no charge. By comparison, the orbiting electrons are negatively charged, and they are much smaller than neutrons and protons. In an electrically neutral atom, the number of protons (positive charges) in a nucleus is exactly equal to the number of electrons (negative charges). The positively charged protons and the negatively charged electrons determine the chemical behavior of an atom. They are also the basis for the "atomic number" given to the elements.

The periodic table lists all the known elements (such as hydrogen and uranium). The atomic number of an element on the periodic table represents the number of protons and electrons in the atom. Hydrogen has an atomic number of one. This indicates that the hydrogen atom has one proton as well as one electron.

The nucleus is held together by the nuclear forces that bind protons to neutrons, neutrons to neutrons, and protons to protons. The number of neutrons in an atom of a particular element can vary. For example, hydrogen can have zero, one, or two neutrons. It has neither more nor less than one proton, however. Chemically, atoms that have the same number of protons but different numbers of neutrons behave in the same way and are called isotopes. In the case of hydrogen, there are three isotopes. The most common isotope of hydrogen, sometimes referred to as protium, has one proton, no neutrons, and one electron. Two other forms of hydrogen—deuterium and tritium—also exist (Figure 1). Deuterium has one proton and one neutron. Tritium has one proton and two neutrons and is radioactive.

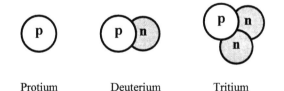

Protium   Deuterium   Tritium

Figure 1   Nuclei of the isotopes of hydrogen. All have one proton (p). Deuterium and tritium have one and two neutrons (n), respectively.

Tritium, deuterium, and protium (or hydrogen) are all hydrogen isotopes. Again, chemically, they behave in a similar manner. The electrons determine the chemical behavior or reactions of each atom.

It is possible to change or transmute one element into another by changing the number of protons and neutrons in the nucleus. Certain elements—especially particular isotopes of elements such as uranium and plutonium—can undergo a process called fission, in which the atoms split in a controlled manner. The result is the production of energy—and this is the basis for nuclear energy. An enormous amount of energy binds the subatomic particles into an element. Breaking these bonds results in energy release, and some of this energy can be harnessed for conversion into a usable form (most often heat) for the generation of electricity.

When an atom undergoes fission, the nucleus of the atom is split into two smaller nuclei of differing elements, resulting (again) in the release of energy. The process can occur spontaneously or be started by the bombardment of the nucleus with neutrons. In a nuclear reactor, a quantity of nuclear fuel is bombarded with neutrons from some neutron source. This bombardment will cause the nuclei to split into other substances and results in the release of additional neutrons and the production of energy (Figure 2).

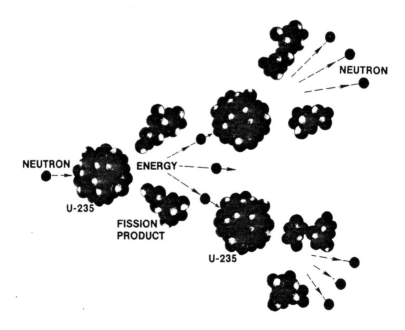

Figure 2   Simplified fission reaction. (Diagram from Department of Energy 1987, 13.)

Three possible outcomes can follow this initial fission reaction. First, the neutrons generated may fail to initiate additional fission reactions, causing the rate of reaction to decrease (eventually to zero). This "subcritical" reaction is not self-sustaining. The second possibility is for the fission rate to increase. It will accelerate both the rate of reaction and the amount of energy produced. This chain reaction is called a "supercritical" reaction. The third possibility is a critical reaction in which the neutrons initiate fission at a steady rate. This last is the normal situation in a reactor; it is self-sustaining until the rate is changed by the operating staff.

To optimize the likelihood of a fission reaction occurring, the neutrons must travel at the right speed. Some reactors, known as thermal reactors, use moderators to slow down the neutrons to ensure the optimal speed. Typical moderators include light water (that is, ordinary water), heavy water (water with a higher than normal percentage of deuterium), graphite, or beryllium. The energy or speed of the neutron needs to be strong enough to overcome the forces of the nucleus.

The neutrons move away from the fission site at high velocities and bump into other atoms. When they do, some are absorbed by these atoms, and the reaction continues.

The reaction produced within the core of a reactor generates heat. This heat must be carried away from the core to prevent overheating and to generate electricity. The way the heat is carried away from the core depends on the type of the reactor. For example, in a water-based reactor, water circulates around the core, heats up, and moves to a heat exchanger and then to a turbine, or it is allowed to boil (with the resulting steam going directly to a turbine). In either case, the turbine powers the generator, which produces electricity that enters the power grid and is distributed along the lines to customers.

Nuclear Fuel

The fuels that are used in reactors are uranium-235 (235U) and plutonium-239 (239Pu). Of these, 235U is the more important. It is a fissile material, which means that it can undergo fission when bombarded with neutrons of any energy. The other isotope of uranium, uranium-238 (238U), is called a fertile material, because it can be transformed into fissile material by neutron bombardment. It is also fissionable, but only by a high-energy neutron.

Uranium-235 is a relatively rare element that occurs naturally. It represents only 0.7% of uranium. Uranium is found in ores located around the world. In the United States, the largest deposits are in the West (mainly in Colorado). Significant commercial deposits are also located in Russia, eastern Europe, Australia, Canada, and Africa. Most uranium mining is surface mining rather than tunnel mining: surface mining is more economical, and radon gas presents some health issues for workers involved in tunnel mining. The extracted ore must be refined to separate the uranium. Next, the 235U must be concentrated, or "enriched," to increase the percentage of 235U present in the uranium. For most commercial reactors, the concentration must be increased from 0.7% to 3–5%.

Fuel production, then, begins with the mining of ore. Once the ore is mined, it is crushed so that the uranium can be extracted. The chemical extraction results in a product that is often yellow (hence the name "yellowcake"). It is then dissolved in nitric acid, filtered, processed, and enriched. Technicians can enrich uranium by turning it into a gas and passing it through filters, separating it by weight in a centrifuge, or using laser excitation. The next step is the creation of uranium oxide, after which the fuel is formed into fuel pellets. These pellets are high-density ceramic fuel pellets that are placed in metal tubes called fuel rods. When a collection of fuel rods is clustered together, it is called a fuel assembly. A number of fuel assemblies are grouped to form the reactor core.

In the normal course of events, fuel has a reactor life of up to six years. A reactor must be refueled periodically to maintain the critical mass of enriched uranium needed to sustain the chain reaction at an efficient level. First, control rods made of a material that absorbs neutrons must be lowered into the reactor core to stop the chain reaction.

The reactor core is contained in a tank called a reactor vessel, which is typically a two-piece structure made of steel. The walls must be thick, because the reactor coolant operates at very high pressures and temperatures. The reactor vessel (or pressure vessel) is filled with water, which acts as a shield. The top part of the vessel, called the head, is unbolted to allow access to the core for refueling. The head is removed by a polar crane (which runs on a track over the reactor vessel) and is moved to the side of the vessel, stored on a support stand, and serviced during the refueling. About one-third of the fuel assemblies are removed and placed into a pool of water for storage, protection, and cooling during refueling, and new fuel assemblies replace the spent fuel. While this is happening, maintenance on the reactor can be done. (At one time, the entire process took weeks to months; now it takes several days. Some reactors can undergo such maintenance without going offline, but TMI is not of that type.)

When the maintenance is complete and the fuel is replaced, the vessel head is lowered into place and the bolts reattached. At this point, the reactor can be started. Neutrons are emitted from the neutron source and the control rods are removed. The core is monitored as it reaches critical state to ensure that everything is working properly.

The removed fuel is physically hot as well as highly radioactive. After a number of years in storage, the radioactivity and heat lessen. The rods can then be put into concrete and steel cylinders called "casks" for eventual final storage. While other countries reprocess their fuel to recover unburned uranium, the United States does not, opting instead for the "once-through cycle": the fuel is destined for storage and eventual disposal after being removed from the reactor.

Nuclear Reactors

Reactors all need a source of energy, coolants, moderators, and a process that converts the heat of radioactivity into usable electricity. The source of energy is the uranium in the fuel rods. The coolant—the substance that carries the heat away from the core to the steam generators—can be water, heavy water, air, helium, liquid sodium, or a liquid sodium–potassium alloy. The result is both energy production and a cooling of the core. As noted above, moderators are materials that slow down the high-energy neutrons so they are at a speed that is more conducive to initiating fission. Water (heavy or light) can serve as both a moderator and a coolant. Other moderators include graphite, liquid sodium, and beryllium.

There are several types of nuclear reactors. Light water reactors (LWR) and heavy water moderated reactors (HWR) are the most common. Other types include high temperature gas-cooled reactors (HTGR), gas-cooled thermal reactors, fast breeder reactors, and a number of advanced light water reactors.

In the United States, only light water reactors are used commercially today. (Some gas-cooled and sodium-cooled reactors were used in the past.) Light water reactors are so named because they use regular water as the coolant. The fuel is uranium, enriched to 3–5% 235U. The fuel pellets are placed in tubes—usually of a zirconium alloy called zircalloy—and each tube is approximately 1 centimeter (.4 inches) in width and 3–4 meters (10-13 feet) long. LWRs can be either pressurized water reactors (PWR) or boiling water reactors (BWR). (See Figures 3 and 4, respectively.)

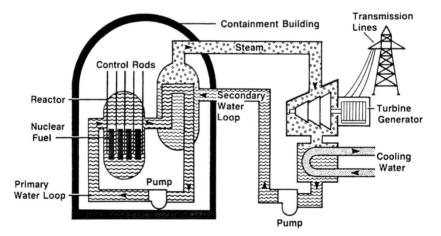

Figure 3   Pressurized water reactor (PWR). (Diagram from Department of Energy 1987, 33.)

Three Mile Island Units 1 and 2 are pressurized water reactors. In PWRs, the vessel is under high pressure and temperature. According to the laws of physics, when water is kept at a high pressure, it can be heated to a higher temperature without its changing to steam. Water in the first loop (the primary loop) is heated by the core to 600°F (315°C). The heated water's energy is transferred to water in the secondary loop. Here the water is converted to steam and flows to the turbine, where it generates electricity for the power grid. The steam then flows through the condenser and returns to the steam generator, where the cycle begins again. The water from the reactor vessel is confined to its loop, separate from the secondary loop. Heat—but no water—is transferred between the primary and secondary loops.

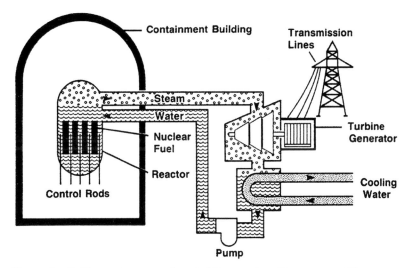

Figure 4   Boiling water reactor (BWR). (Diagram from Department of Energy 1987, 32.)

Boiling water reactors are so called because they utilize boiling water that is heated in the core of the reactor. There is no steam generator in a boiling water reactor, so the resulting steam flows directly to the turbine. The pressure in the reactor vessel is lower than in the PWR, but it is still maintained at a relatively high pressure so that the steam can be heated to 545°F (285°C). The steam flows from the reactor vessel directly to the turbine and then to the condenser, where it cools and is pumped back to the reactor vessel—and the cycle continues.

Among heavy water reactors, a common type used in Canada is the CANDU reactor (CANada Deuterium Uranium). It uses natural uranium and heavy water and is capable of being refueled while online. Gas-cooled thermal reactors use carbon dioxide as the coolant; they have high thermal efficiency and are popular in the United Kingdom. The high temperature gas-cooled reactor (a variation on the gas-cooled thermal reactor) utilizes helium as the coolant.

Several types of breeder reactors have also been developed. These include the liquid metal–cooled fast breeder, gas-cooled fast breeder, molten salt breeder, and light water breeder reactors. Breeder reactors start with a fissile material, such as 235U, that initiates a chain reaction. The neutrons produced react with uranium-238 to produce 239U, which decays to form the intermediate neptunium-239 and then plutonium-239, another fissile material. These are called breeder reactors because they create more fissile fuel at the end of the cycle than at the beginning, which lessens the need for uranium mining and refining. One of the concerns associated with breeder reactors, however, is the weapons potential of the resulting plutonium, which makes this research program controversial. Still, several countries (generally those with limited uranium resources) are pursuing this type of reactor.

Advanced light water reactors (ALWR) are based on LWR technology. Their newer technologies result in more simplicity, redundant and passive safety features, shorter construction timelines, and financial savings. There are a number of designs, including the advanced boiling water reactor (ABWR); the System 80+, a PWR design; the AP600/1000 PWR, which utilizes a number of passive cooling techniques; and the simplified boiling water reactor (ESBWR), which utilizes the natural circulation of the coolant as well as a rapid depressurization capability. In many cases, these improvements were initiated to simplify systems, lower costs, and reduce the likelihood of an accident such as the one that occurred at TMI.

Viewed from the outside, all of these nuclear plants have several common features. One of the most distinctive is the cooling tower. This concrete cylindrical structure rises high above the rest of the facility and serves as a heat exchanger. Heated water flows into the tower, where it is cooled and pumped back into the reactor building, providing cooling water for condensing the steam used to turn the turbine. A dry cooling tower lowers the temperature of the water by evaporating and cooling the water with a flow of air. A wet cooling tower uses a flow of water in addition to air to cool the water in the loop. Each is easily recognizable by the white plume above the tower caused by the vapor resulting from the cooling process. The cooling tower dissipates what is called waste heat. The laws of thermodynamics dictate that any power plant will operate at less than 100% efficiency and will thus produce a certain amount of waste heat. Several factors preclude using this heat source on a widespread basis for home heating or industrial processes. First, the need for heat is seasonal in most areas, so it could only be used at certain times of the year. In addition, the logistics of getting the heat to the right locations are complex—and vast amounts of heat are generated, more than many areas could practically use. Still, a few plants around the world do harness waste heat. The process of generating electricity and utilizing the waste heat is called cogeneration.

Another common feature of all nuclear power plants is the containment building, which houses the reactor core. It is made of reinforced concrete and steel and typically looks a bit like a large, thick farm silo. The condenser, turbine, and generator are located in a building near the reactor building and contain the equipment needed to convert the heated water or steam to electricity. And in a control room external to the reactor, technicians monitor the activities in the core and in other equipment.

Safety

Nuclear reactors are designed and built with safety as an integral feature. The systems are redundant, so if something does go wrong, a backup system will still operate. Such features include the fuel design, the containment building, and the numerous safety systems.

The containment building, as noted above, is constructed of rein-forced concrete and steel; it is meant to withstand a crash into the building, natural disasters, and the extreme heat and pressures of nor-mal operation and accidents. The design of the TMI reactors even took into account the possibility of a plane crashing into them. (This was actually of practical import at Three Mile Island, as the Harrisburg airport is only a short distance from the island.) Within the reactor, the fuel pellets are contained in cladding—that is, tubes made of protective materials such as zirconium, stainless steel, or aluminum. The cladding prevents the leaking of radioactivity and the corrosion of the fuel rods.

The control rods constitute a major safety feature of nuclear reac-tors. Made of materials such as boron or hafnium, the rods are located with the fuel in the core and can quench or absorb the neutrons that maintain the chain reaction in the core. By absorbing the neutrons, the rods can slow or stop a chain reaction. In BWRs, the control rods are moved as the properties of the fuel change over time—in essence, fine-tuning the reactor. In PWRs, boric acid (a chemical) is added to the coolant to allow for changes in the reactivity of the fuel as it is used. When the reaction is started in the core, the control rods are removed from the core and the neutrons from various sources bombard the fuel to begin the reaction. The rods, however, can also be reinserted to trap the neutrons and stop the reaction. Rods can be controlled from the control room, but if the systems fail or if there is a mechanical prob-lem, the control rods can drop by gravity into the reactor core. This process can be initiated either manually (by the operator) or automati-cally (by the reactor protection system).

Removing the energy generated by the radioactive decay of the nuclei formed in the fission process is of the utmost importance. These nuclei, called fission fragments, are highly radioactive and emit energy. The energy eventually appears as heat ("decay heat"). Even though the fission process is stopped by the insertion of the control rods, the decay of the fission fragments continues. The amount of energy released is considerable and can amount to as much as 6% of the energy produced by the reactor. The reactor would be damaged if the decay heat were not removed. To ensure its removal, a number of safety systems are available. These systems, called the emergency core cooling and the decay heat removal systems, provide cooling water to the reactor in emergency and normal operations, respectively.

Safety, of course, extends to the facility employees, who undergo training, testing, and periodic review. While this was true before the TMI accident, the process changed after the accident and is explained in a later chapter.

History of the Nuclear Industry Through 1979

The history of the nuclear field might well be said to have begun with Roentgen and the discovery of X rays, Madame Curie and her discovery of radium, and Albert Einstein with his special theory of relativity. From a practical standpoint, however, it begins with the Manhattan Project during World War II. The power of the atom was harnessed for weapons during this intense time of study and application, a time that brought together some of the greatest minds in science and technology. The goal of the Manhattan Project was to build an atomic bomb, because there was fear that Germany was developing such a weapon.

The peaceful use of the atom became a major interest after the war. In 1953, the Eisenhower administration proposed an "Atoms for Peace" program. The first experimental reactor was brought online in 1951; in 1957, only a dozen years after the end of the war, President Eisenhower commissioned the first full-scale nuclear power plant in Shippingport, Pennsylvania, located west of Pittsburgh. The plant was a pressurized water reactor jointly owned by the Duquesne Light Company and the U.S. Atomic Energy Commission (AEC). (The AEC was combined with other energy-related government agencies into the Energy Research and Development Administration—and eventually into the U.S. Department of Energy.) The Shippingport Atomic Power Station officially came online on December 18, 1957, and it produced power for the Pittsburgh area until 1982, when it was shut down. It was decommissioned in 1985, which gave it the dual distinction of being the first commercial plant to go online and the first to be decommissioned.

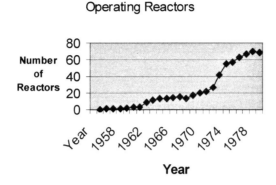

Figure 5   Number of reactors in the United States through 1979. (Data from Energy Information Administration, "Annual Energy Review 2002"; see http://www.eia.doe.gov/emeu/aer/pdf/pages/sec93.pdf.)

Commercial plants followed on a regular basis soon after Shipping-port (Figure 5). In 1959, Dresden Unit 1 in Illinois (the first privately financed reactor) came online. A number of reactors were subsequently planned; in 1973 alone, forty-one plants were on order. Indeed, nuclear power plants accounted for 4.5% of the electricity generated in the United States in 1973. By 1979, that figure had more than doubled to 11.3%. The use of nuclear energy as a power source had grown greatly in twenty years. Meanwhile, nuclear technology was being utilized in naval and commercial vessels, scientific stations in remote areas, and the space program. In less than thirty-five years, nuclear technology had moved from secret military research to practical productivity.

Table 1   Sources of electrical power generation in the United States, 1959–1979

	Fuel sources and percentage of total production						Total energy
Year	Coal	Natural gas	Crude oil	Nuclear	Hydro-power	Other	(billion kilowatt hours)
1959	53.042%	20.549%	6.560%	0.028%	19.793%	0.028%	713.4
1964	53.302	22.285	5.774	0.334	18.264	0.030	987.2
1969	48.841	23.058	9.533	0.962	17.537	0.062	1,445.5
1974	44.292	17.115	16.088	6.095	16.265	0.150	1,870.3
1979	47.763	14.640	13.485	11.339	12.578	0.196	2,250.7

SOURCE: Data from Energy Information Administration, "Annual Energy Review 2002," Table 8.2a; see http://www.eia.doe.gov/emeu/aer/txt/ptb0802a.html.

Three Mile Island

Few people realize that Unit 2 of Three Mile Island was originally slated for construction as the Oyster Creek Nuclear Generating Station Unit 2 in the Forked River area of New Jersey, not in Pennsylvania. The New York metropolitan area had placed increasing demands on the power grid, and a new reactor was one of the solutions to this need for more electricity. After the plans were completed, however, labor problems forced the company to consider other alternatives for the plant. To avoid the same labor issues, the plant would have to be built outside of New Jersey. The island in the Susquehanna proved to be a viable alternative for several reasons. Another reactor was already being built on the island, so the geologic surveys and other preliminary work had been completed. The community was already aware of the island's use as the home of a nuclear reactor. In addition, both reactors had been designed by Babcock and Wilcox and were to be built by United Engineers and Constructors. Only the design engineers were different. Unit 1 was using Gilbert Associates; the newly moved Unit 2 would use Burns and Roe.

A number of practical issues had to be addressed in moving the reactor to Pennsylvania. For instance, the Oyster Creek reactor was designed to use saltwater to cool the steam that drives the turbines, but TMI is on the Susquehanna, a freshwater river. After studying the problem, engineers decided that the water purification system originally planned for Oyster Creek would be more than adequate for TMI. Also of concern was the control room. Facilities with more than one reactor often share a single control room. In the case of the two reactors on Three Mile Island, the decision was made to keep the control rooms separate and not to modify the second to conform to the one at Unit 1. Technical staff could be certified in one or both of the control rooms, but they would require separate training and certification.

Construction of Unit 1 began in 1967. The unit started (or "went critical") in early 1974, and commercial service began on September 2, 1974. At the time of the accident at Unit 2, Unit 1 was down for refueling.

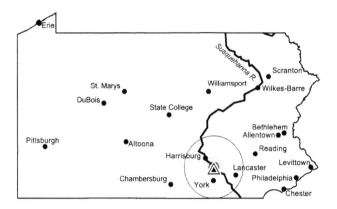

Figure 6   Location of the Three Mile Island Nuclear Generating Station. (Map courtesy of Derrick Beckner, University Libraries, The Pennsylvania State University.)

Construction of Unit 2 began in 1970, the fuel was loaded, and testing began on March 28, 1978, a year before the accident. Testing had only been in progress for one day when, on March 29, the reactor shut down. A pressure-regulating valve had opened and stayed partially open, allowing coolant to leave the system; this resulted in the reactor's pressure dropping. Over the next nine months, the system was tested at least fifteen times as part of the plant's test program. A number of problems were uncovered, especially ones involving the valves. On December 30, 1978, TMI Unit 2 was declared commercial—and began to generate electricity for its customers.

Reactor Accidents

The accident at TMI Unit 2 was not the first accident at an electrical generation plant, nor was it the first nuclear reactor problem. (In fact, there are certain dangers inherent in all forms of energy production, as will be discussed below.)

In 1957, a serious incident occurred at Windscale, an advanced gas-cooled reactor located in Cumbria in the northwestern part of England.

One of the main goals of the Windscale reactor was to produce weapons-grade plutonium for the British, but it also supplied some electrical energy. Construction began in 1947. Pile 1 went critical in October 1950; Pile 2 went critical in 1951. ("Pile" is a term used for early reactors.) Windscale used graphite as the moderator, and graphite would also provide structure to the core elements.

On October 8, 1957, one of the physicists at Windscale was working with Pile 1 on a routine procedure to release energy that could be trapped in the core during neutron bombardment. This was done by raising and lowering core temperature to try to realign the graphite block, which had become misshapen. As the physicist pulled out boron control rods, a fuel rod ignited. The fire burned for several days. It was finally put out by flooding the core with more than 1.32 million gallons (5 million liters) of water delivered by four fire hoses over a period of thirty hours. Radioactive iodine-131 was released. For public safety, the government ordered the milk produced from the surrounding area to be dumped. Both the damaged Pile 1 and Pile 2 were shut down. No one was killed in the accident, but many sources have listed it as the worst accident until Chernobyl. The first phase of Pile 1's decommissioning was completed in 1999.

Another accident occurred at the Brown's Ferry Nuclear Power Plant located in Decatur, Alabama. The three units that make up Brown's Ferry—all boiling water reactors—are part of the Tennessee Valley Authority. They came online between 1973 and 1977. On March 22, 1975, an electrician was checking for air leaks in an area under the control room shared by Units 1 and 2. He used a candle as a means of locating the air flow. The flame started a fire in the insulation, which lasted seven and one-half hours and destroyed more than a thousand control cables, some of which were involved with the operation of safety devices. There was no release of radioactivity into the environment, and Unit 2 was shut down immediately, but the damage was such that the reaction in Unit 1 could not be shut down quickly. The ability to shut down a reactor rapidly is an important safety feature. More than 1,600 cables and circuits were damaged, so costly repairs were needed. Unit 2 was restarted in May 1991.

The most infamous—and by far the most devastating—reactor accident occurred at Chernobyl in Soviet Ukraine in 1986. Chernobyl was a light water–cooled, graphite-moderated reactor that began producing power in 1983. Like many Soviet reactors, it did not have the reinforced containment building commonly used for reactors in other countries. On April 25, 1986, a test was run on Unit 4. The purpose of the test was to see what would happen if power was terminated. During the test, the emergency cooling system and the automatic control rod system were turned off. Warning signals were bypassed. Heat and steam built up, and two explosions ripped through the reactor. The first explosion damaged the core and the structures. The second one blew the top off the reactor building. The graphite core reached temperatures over 5,000°F (2,760°C) and burned. Radioactivity was released for days.

Headlines around the world announced the explosion and contamination. Thirty-one people died within a short time of the accident. There are a number of reports on the long-term effects of Chernobyl on the local population as well as on other areas that received some contamination. They range from statistically few health abnormalities to large numbers of deaths and projected massive numbers of cancers in the future. The most extensive studies have been done by the Organisation for Economic Co-operation and Development Nuclear Energy Agency, the International Atomic Energy Agency, and various independent researchers. The one finding that seems to be consistent is the increased number of cases of thyroid cancers and other thyroid abnormalities, especially in children. On this last issue, it has been stated that "the most likely number of cases over all time in all affected countries has been estimated at 4400, with 7800 as an upper bound" (Thomas and Zwissler 2003, 210). The United Nations Scientific Committee on the Effects of Atomic Radiation (UNSCEAR) found that there is no evidence indicating other fatal diseases and that "the death toll attributable to the Chernobyl accident could turn out to be on the order of a thousand" (ibid.). This estimate contrasts greatly with the much higher death tolls projected by Greenpeace and the national governments of the states surrounding the reactor.

A number of issues have made reliable studies difficult. After the accident, people from surrounding areas were evacuated and dispersed, making follow-up more complicated. The radiation from this accident covered a very large area and a number of countries, adding to the problems of study. The lack of health data in the Chernobyl area prior to the accident precludes comparison data. In addition, health effects often take many years to appear, so conclusive results will not be known for some time. Concern about the contamination of the land and wildlife remains. Tests and studies will continue in the area to monitor long-term effects.

Summary

The second half of the twentieth century saw major advances in nuclear technology. After World War II and the Manhattan Project, scientists turned their attention to harnessing nuclear energy for peaceful uses. Reactors were developed to address the growing needs of industry and residential consumers. Nuclear energy was adapted to power remote research stations, space probes, and naval vessels. With growing energy demands, increasing concern about environmental pollution, and dependence on imported fossil fuels, nuclear energy promised to be an effective technology that would address energy needs in a cost-effective manner. But not everyone embraced this promise. Having learned lessons from the political, civil, and environmental protests of the 1960s, groups organized to protest the use of nuclear technology. Even people who were not politically active were concerned about a technology that they associated with the development of the atomic bomb. There was fear of nuclear technology in general; there was fear of nuclear technology coming to the neighborhood. As reactors were built and planned for communities across the country, the acronym NIMBY—meaning "not in my backyard"—became a reality for many. Even Hollywood entered the discussion with the timely movie *The China Syndrome*.

So while nuclear energy seemed a logical solution to the problems of dependence on foreign oil, the environmental effects of power generation, and the growing need for more electricity, concerns about the safety as well as the health and environmental implications of nuclear plants created hurdles that threatened the future viability of using nuclear technology on a wide scale. These concerns only grew after several accidents at nuclear plants.

2

The Accident

Around 4:00 AM on Wednesday, March 28, 1979, a sequence of events began at Unit 2 of the Three Mile Island Nuclear Generating Station that would have an enormous impact on the future of nuclear power throughout the world. This chapter reviews that sequence of events, beginning with the first signs of trouble that morning through the stabilization of the reactor several days later, and it follows the actions of the key participants—reactor operators, supervisors, the Nuclear Regulatory Commission (NRC), and state and local officials.

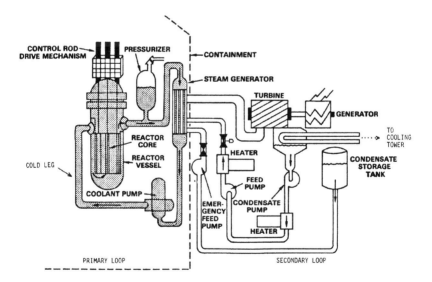

Figure 7 Schematic of TMI Unit 2. (Diagram from Nuclear Safety Analysis Center 1981, 3-2.)

The main feedwater system that supplies water to the steam generators (see Figure 7) malfunctioned and shut off the flow at 4:00 AM. Water must be fed continually to the steam generators to make up for the flow of steam from the steam generators to the turbine. Should the flow of cooling water be interrupted, the reactor could overheat.

While the exact cause of the failure of the feedwater system is still unknown, maintenance was being performed on the condensate polishing system at the time of the accident. The polishing system is used to purify the water in the feedwater system. This maintenance may have caused a fault to occur, shutting down the feedwater system, but it is not clear that this is what happened.

Immediately after the feedwater failure, alarms began to sound. Within minutes, more than one hundred alarms were sounding in the control room. An auxiliary feedwater system should have started automatically but did not. The purpose of this system is to provide an emergency source of cooling water to the steam generators in just such an event, but due to a maintenance error after a test of the system, critical valves in the system were left closed (in violation of NRC regulations). Without this critical water supply, the boiling water in the steam generators would boil away completely, resulting in a rapid rise in the reactor's cooling system temperatures and pressures.

The turbine shut itself down automatically. Within seconds, the reactor's control system shut down the reactor by dropping the control rods, thus stopping the fission process. As temperatures and pressures continued to rise in the reactor system, water flowed into the pressurizer. The pressurizer is normally about half-filled with water and half-filled with steam, providing an expansion tank to cushion the reactor system against changes in pressure during normal operation. A power-operated relief valve also helps control pressure during abnormal events.

The pressure-regulating valve—the same one that had malfunctioned during the initial testing of the reactor—opened to reduce the pressure in the reactor and associated reactor systems. Steam began to flow from the valve through piping into a collecting tank located in the basement of the reactor containment building. The system was designed to trap any flow from the valve because it could be contaminated with radioactive material.

The flow from the reactor system caused the pressure to decrease, as expected. What happened next was not expected—and it triggered the chain of events that would make TMI a major incident in the history of nuclear power. After the pressure had decreased, the pressure-regulating valve should have closed, but once again, it failed. The valve stuck before it could close completely, and operators had no way to monitor the valve's precise position. A backup system that measured the temperature of the piping between the tank and the s could have helped operators determine whether the valve was open, but small leaks from the valve during the previous month's operation caused the temperature to remain high, masking the continued flow. The operators were thus unaware of the catastrophe in the making. They did not know that vital cooling water was flowing from the reactor out through the valve to the basement's collecting tank. So much water was lost, in fact, that the tank eventually overflowed; water spilled onto the basement floor of the containment building and was pumped to storage tanks in the adjacent auxiliary building outside the containment building.

As water leaked through the open valve, pressure in the reactor's cooling system dropped. The emergency core cooling system, which was designed to provide a source of cooling water in the event of this type of occurrence, started automatically once the pressure had dropped too low. Again, the backup systems were operating as expected, but then something else happened that misled the operators and eventually caused severe damage to the reactor.

Drawing upon their years of training and the procedure manuals, the operators were attempting to assess what was happening, all the while being bombarded by more than a hundred alarms and flashing lights. Of great concern was the threat of overfilling the reactor system: this could be catastrophic, because it could lead to the system's becoming "solid." In such a case, the reactor coolant system would completely fill with water, over-pressurizing the reactor cooling system and causing it to rupture and fail.

The water level and volume in the reactor system were not measured directly. Instead, operators relied on a measurement of the water
level in the system's pressurizer. As noted above, the pressurizer is
normally half-filled with water and half-filled with steam. The
operators saw that the level of the water in the pressurizer was
increasing despite the loss of water from the reactor system. The
increasing water level misled the operators: they thought that there was
adequate water inventory in the primary cooling system, which in turn
led them to conclude that the system might go "solid." They therefore
turned off one of the two pumps in the emergency core cooling system,
thus reducing the flow of water critical to reactor cooling. Shortly
afterward, still believing that there was adequate water in the primary
cooling system, the operators shut off the remaining emergency cooling system pump. They did not realize that this was a loss-of-coolant
accident. Their interpretation of the data determined their corrective
actions. Had the reactor operators realized what was happening and let
the emergency core cooling system perform its function, it is likely
that the reactor would not have been damaged. Instead, the only problem would have been the overflow of slightly contaminated water into
the containment building sump.

Over the next one and one-half hours, coolant continued to flow
from the partially opened valve, and the water level in the primary
cooling system continued to drop. Because of the decreasing water
level, the pressure in the cooling system also fell. By 5:30 AM, the
pressures had dropped so low that the large reactor coolant pumps used
to circulate the water through the reactor and the primary system began
to vibrate. If the pumps had been allowed to continue working, the
vibration would have destroyed them. The operators began to shut
down the pumps to reduce the possibility of damage.

At this point, the operators thought that they might have a leak—but
not in the reactor cooling system. Rather, they suspected that there was
a leak in the steam system leading from the steam generators to the
large valves designed to isolate the steam generators from the turbines.
Such an event is called a main steam line break accident, which has
many of the features that the operators were noting. Unfortunately, the
operators missed a key observation: the temperature and pressure in

the reactor system were converging to a common state. The pressure at which water will boil for a given temperature is referred to as the saturation pressure (at 212°F—or 100°C—the saturation pressure is 1 atmosphere). By now, the coolant pressure in the primary system had reached the saturation pressure for the coolant temperature. Had the operators realized this, they might have concluded that there was a leak in the primary system, not in the steam system.

Meanwhile, water continued to spill onto the floor of the containment building. From there, it was still being pumped into tanks in the auxiliary building—and they had become full as well. Water began pouring onto the floor of the auxiliary building, and radioactive gases began to find their way from the cooling water through the auxiliary building ventilation system out to the outside world.

The coolant level dropped to the top of the reactor core, and the fuel rods' temperature began to rise. Despite the shutdown of the fission process at the beginning of the accident, energy was still being released in the fuel by the decay of the fission products generated during the operation of the reactor. Without cooling, the rods would eventually burst and possibly melt, releasing large quantities of radioactivity into the cooling system. The operators were concerned about such an event, but they were distracted and failed to realize what was happening until fuel damage began to occur. Even at this point, if they had realized that they were dealing with a loss-of-coolant accident, the core could have been saved with only minor damage (if any). The operators only needed to turn on the emergency core cooling system to full flow, thus refilling the reactor and cooling the core.

When fuel overheats, the first result is the rapid oxidation of the zirconium alloy from which the fuel rod tubes are made. The zirconium reaches a critical point and oxidizes very rapidly, releasing heat that further accelerates the oxidation. The process also releases hydrogen from the water. Temperatures in the top of the core reached this point not too long after the water level dropped below the top of the fuel in the reactor. Events were now arriving at a critical point. If the level continued to drop, the core's temperature would rise to the fuel's melting point; if that were to happen, no one knew for certain whether the damaged fuel could then be cooled.

Over the next hour, the water levels in the core dropped further, exposing more and more fuel to damaging temperatures. As the zirconium heated up, oxidized, and failed, the fuel began to overheat and melt. Meanwhile, radioactive gases normally contained in the rods were released into the cooling system and then passed out the stuck-open valve to the reactor building. High-radiation alarms began to sound, indicating the presence of abnormal radiation levels in the containment building. The pump in the reactor building's basement continued to pump the water spilling from the collecting tank into the auxiliary building. This water was now highly contaminated with radioactivity from the fuel, and this radioactivity was being released into the environment by way of the auxiliary building's ventilation system. Radiation alarms began to sound at many points throughout the plant.

Around 6:22 AM, the operators realized that the power-operated relief valve was stuck open. They closed a backup valve, stopping the flow of water from the reactor. Because of the high coolant temperatures, normal cooling of the fuel could not be restored. In addition, the hydrogen generated by the zirconium's oxidation was interfering with the heat transfer in the steam generators. By 6:55 AM, a site emergency had been declared.

By 7:00 AM, a team had assembled at the site to assist the operators. The TMI personnel were already following the site emergency plan and had notified the Pennsylvania Emergency Management Agency (PEMA), the Pennsylvania Bureau of Radiation Protection (BRP), and the U.S. Department of Energy's Radiological Assistance Plan Office. (The NRC was not notified until nearly 8:00, however, as a result of poor communication protocols on its part: the local NRC office was not manned, and telephone calls were answered by an answering machine until 8:00 AM, when the normal workday began.)

At 7:24 AM, in recognition of the extremely high levels of radiation in the containment building, the site emergency was elevated to the status of a general site emergency. Teams of technicians were dispatched to points on the island and to the nearby town of Goldsboro to check radiation levels. Although radiation had been released from the reactor, the levels were found to be extremely low or not detectable. By 8:00 AM, TMI personnel had concluded that there had been some fuel damage. (The extent of the damage would be considerably underestimated until a "quick look" into the damaged reactor was made several years later.)

At 8:25 AM, news of the TMI accident began to hit the airwaves. A local top-40 radio station, WKBO, broke the news first. By 10:00 AM, the story had spread across the nation, with broadcasts on the major radio networks. The team that had been sent to Goldsboro erroneously reported that it had detected radioactive iodine. Iodine is of particular concern because it is attracted to the thyroid gland and can cause thyroid cancer. This report was repeated at a news conference by representatives of the Pennsylvania BRP. Meanwhile, Metropolitan Edison—one of the power companies that owned and operated TMI— declared that no off-site readings had been found, which led to confusion and the beginning of the press's distrust of Metropolitan Edison.

Around 9:00 AM, TMI personnel discovered the water in the auxiliary building and stopped the pumping from the containment building's basement. This slowed the release of radioactivity from the plant. The most serious problem then facing the operators was getting cooling water back into the reactor to cool the core. The emergency core cooling system was finally turned on, which initiated a process of injecting cooling water into the primary system and then discharging it through a relief valve. By noon, press representatives from around the world began to arrive. Lieutenant Governor William Scranton held press conferences to discuss the TMI situation. (Although lines of communication were open between his office and Metropolitan Edison, information did not flow in a timely manner, and the lieutenant governor consequently made statements that later were shown to be wrong at the time they were made.)

While operators were restoring the cooling system, the relief valve continued to release hydrogen into the containment building's atmosphere. The hydrogen mixed with the oxygen in the atmosphere, and this mixture ignited at about 1:50 PM. An overpressure of twenty-eight pounds per square inch was reached when the conflagration occurred, but because this was well below the pressure the containment building was designed to take, there was no danger that the building would be breached.

By that evening, local and national television stations featured the accident on their nightly news broadcasts. Eventually the reactor system was cooled to the point where the reactor coolant pumps could be turned on and the normal process of heat removal resumed. This happened at around 8:00 PM—sixteen hours after the accident had begun. No one knew for sure what had happened. Had the reactor been damaged? If so, how extensive was the damage? It is now known that if the emergency core cooling system had been restarted even as late as one and one-half hours after the onset of the accident, there would have been minimal damage to the reactor core.

By Thursday, March 29, the situation appeared to have been stabilized. Radiation monitors in and around the plant, as well as across the river to the west, detected very low levels of radiation. No iodine was detected. But at 2:10 PM, a helicopter sampling radiation levels over the plant measured a burst of radiation at the plant's vent, which is located along the side of the containment building. This caused little or no concern among officials. A second release the following morning, however, worsened the relationship between the NRC and the state—a relationship already frayed by miscommunication and misunderstandings.

At 7:56 on Friday morning, radiation was released from storage tanks inside TMI as workers sought to relieve the pressure that had built up in the tanks. A helicopter monitoring the radiation levels overhead recorded a level of 1,200 millirem per hour, which is a level many times the normal reading. This report found its way into a meeting of NRC officials, but unfortunately, it was erroneously assumed to represent a ground-level recording. In fact, it had been recorded 130 feet above the plant and posed no serious danger to the population, because no one would be exposed to this level of radiation on the ground. The error resulted in a recommendation from the NRC that state officials order an evacuation of the population within ten miles of the plant. Confusion reigned over the next several hours, eventually culminating in a press conference by Governor Richard Thornburgh at 12:30 PM. Thornburgh stated,

Based on advice of the Chairman of the Nuclear Regulatory Commission and in the interest of taking every precaution, I am advising those who may be particularly susceptible to the effects of radiation, that is, pregnant women and pre-school age children, to leave the area within a five-mile radius of the Three Mile Island facility until further notice. (Cantelon and Williams 1982, 54)

As a result of this advisory, more than 3,500 pregnant women and preschool-aged children left the area.

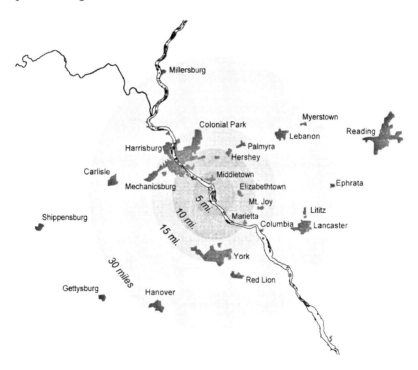

Figure 8 Map of the area surrounding Three Mile Island. (Map courtesy of Derrick Beckner, University Libraries, The Pennsylvania State University.)

Meanwhile, the presence of hydrogen in the primary cooling system continued to cause concern throughout the day on Friday. About sixteen hours into the accident, stable cooling had been achieved, but the operators and engineers concluded that the hydrogen had collected in a bubble in the top of the reactor vessel. This fact was widely reported in the media and created a crisis of public fear on what is now referred to as "Black Friday." Experts disagreed on whether a hydrogen bubble could explode. If it did explode, some feared that the explosion would rupture the vessel, causing the core to become uncovered and to over-heat again—with no way to cool it. But the amount of oxygen in the system was far too low to cause an explosion. In fact, during normal operation, hydrogen is *added* to the system to gather oxygen and pre-vent the corrosion of the cooling system's piping.

On April 1, President Jimmy Carter and his wife, Rosalynn, arrived at an Air National Guard airport near TMI. They joined Harold Denton, the NRC's chief spokesperson, and Governor Thornburgh for a tour of the plant and an update on the status of the emergency. The appearance of the President at the accident site was reassuring to most (if not all) residents of the area. President Carter was a trained engineer; he had served several years in the Navy and had experience with nuclear submarines. He understood the technology. His visit sent the message that the situation was under control. On April 4, the reactor was stabi-lized, and the accident was over.

Summary

The events of March 28 through April 4, 1979, were unprecedented in U.S. nuclear history. The industry, government, media, and the general population were caught off guard and had to scramble to understand the technological problems and to determine the correct actions to take to control the reactor and to protect the local populace. Unfortunately, mistrust and confusion resulted, further complicating the situation.

The accident at TMI resulted in extensive damage to the reactor core. Nearly all the fuel rods were damaged. Much of the fuel melted, forming a mass in the lower part of the core region (see Plates 15 and 16). The molten mass eventually burned through the lower core sup-port plate, nearly filling the lower head of the vessel. Although no one knows for sure what would have happened if the leak had continued, it is quite possible that in another twenty minutes or so, the lower portion of the reactor vessel would have melted, releasing hot molten fuel onto the floor of the containment building and posing an even greater, possibly uncontrollable release of radiation into the environment.

The amount of radiation released into the environment was approximately thirty-four curies of iodine and ten million curies of noble gases, mainly xenon. Most of the radioactive iodine was contained within the primary system. The amount released at TMI was insignificant, because the release occurred in an aqueous environment (unlike the release at Chernobyl). No detectable amounts of the longer-lived radioisotopes of cesium or strontium were found outside the plant. Large amounts, however, were discovered in the primary system, the containment building, and the tanks in the auxiliary building.

It is interesting to note that in September 1977, another Babcock and Wilcox–designed reactor, Davis-Besse in Ohio, had experienced a similar problem involving a rising pressurizer level and falling pressure. Fortunately, there was no damage to the fuel or reactor, but a Babcock and Wilcox senior engineer issued a serious warning that more precise instructions were needed as a result of the incident. In fact, a report from an engineer at the Tennessee Valley Authority—another nuclear operator—questioned whether most control room personnel were adequately trained to respond to such an incident. The letter was sent to Babcock and Wilcox in April 1978—nearly a year *before* the TMI accident.

Power-operated relief valves had stuck open at least nine times in other Babcock and Wilcox–designed reactors prior to the TMI accident. The company neither informed customers of these failures nor upgraded its training program. Consequently, operators were largely unaware of the potential for a loss-of-coolant accident. While the TMI operators performed above the national average on NRC exams, the NRC standards were not as rigorous as they needed to be. The standards allowed conceptually weak training programs to be used with no formal qualification requirements for the instructors.

Numerous studies have since determined the principal causes of the accident. The Kemeny Commission and Rogovin reports are the most notable. They concluded that the accident resulted from a number of factors, including poor training, inadequate equipment design, poor control room design, and (most of all) the industry's attitude, or "mind-set." This mind-set was most evident in the industry's belief that it had designed accident-proof plants and that a serious accident was improbable, thanks to the many safety features engineered into each nuclear reactor.

In the next chapter, we turn to the cleanup effort, which lasted over ten years and cost approximately one billion dollars.

3

The Cleanup of TMI Unit 2

By April 4, 1979, the accident was over. The reactor system had been stabilized and the hydrogen and other non-condensable gases removed, enabling the reactor to cool without interruption. The next step was to bring the reactor to what is referred to as "cold shutdown." For this to happen, the water temperature throughout the cooling system must be maintained at well below 212°F (100°C)—the normal boiling point of water at atmospheric pressure.

Cold shutdown was achieved at Unit 2 on April 27, when the temperatures throughout the reactor's cooling system decreased to below 188°F (87°C). Operators accomplished this through the normal process of generating steam in the reactor. The steam generators were operated over a period of six days, from April 14 to April 19, to lower the temperature to close to cold shutdown. On April 19, the main turbine was engaged to increase steam flow further and reduce the temperature to 188°F.

By April 27, therefore, TMI was out of danger—but this was by no means the end of the story. Now prolonged discussions began over what to do with the damaged reactor. How severely was it damaged? Could it be restored? How much would the cleanup cost, and who would pay for it? Such a task had never before been performed on this scale. Even the accident at Windscale in the 1950s was still not completely cleaned up. Planning at TMI was painstakingly slow, with alternative after alternative being considered. Nearly fourteen months later, in June 1980, the cleanup would finally begin. And TMI would once again vault to the front page of the newspapers.

Before any work could begin, the containment at TMI needed to be vented. As a result of the accident, the reactor building held sixty-four thousand curies of krypton-85, a radioactive form of krypton. The radiation level was thus too high for workers to proceed with the necessary examinations and inspections of the reactor. In November 1979,

Metropolitan Edison requested approval from the NRC for a plan to vent the krypton-85 into the atmosphere. Radioactive krypton is a noble gas and is not a major health hazard (noble gases are not retained in the body). The public reacted swiftly, however. In early 1980 a series of meetings was held. The first two were held near TMI—the first at the Liberty Fire Hall in Middletown, the second at the high school in Elizabethtown. The third meeting was held at the NRC in Washington, D.C.

In the first two meetings, plans for the venting were reviewed and public comment heard. Prompted by the distrust that had built up during the accident, the public mood was confrontational and developed into an ugly scene. The *Philadelphia Inquirer* and other papers reported that citizen after citizen denounced the plans to vent the containment and expressed fear for their safety should the plan be approved and implemented. At each meeting, the residents attacked Metropolitan Edison, the NRC, the Department of Energy, and state and federal representatives, accusing them of endangering their families, lying, and being co-opted by the nuclear industry.

More than five hundred angry residents attended the Middletown meeting. At one point, NRC officials in attendance were referred to as "you animals." The mood of the meeting deteriorated to the point where the state police escorted the government and utility speakers out of the firehouse for their protection. Afterwards, NRC spokesperson Harold Denton said that the meeting was "probably the most raucous meeting we've ever attended."

The meeting in Elizabethtown was only slightly better. A Metropolitan Edison official was belittled by a speaker who said, "How in God's name do you people sleep at night?" Another resident asked, "We're talking about people with hearts and souls who are being traumatized. What do you do with these casualties?" Clearly, fear had taken hold within the population around Three Mile Island despite assurances from the government and industry representatives that venting was the safest way to dispose of the krypton gas trapped inside the Unit 2 containment building.

At the meeting in Washington, six neighbors of TMI told of people driven to the edge by fear and of families leaving the community. They spoke of strange events purported to have occurred in and around the island after the accident and of animals dying and plants withering in mysterious ways. One speaker said, "The people of the State of Pennsylvania feel we've been sold down the tubes by everyone." Psychologists who analyzed the situation felt that there was now an absolute distrust of the government at all levels. To some, the situation was explosive and had the potential for violence (Nuclear Regulatory Commission 1980a, 2:177–79).

A local antinuclear group called PANE (People Against Nuclear Energy) sought a court injunction to prevent the venting. Numerous other alternatives were offered, including a congressional plan. On June 12, 1980, the NRC approved the plan to vent the containment. The appeal to halt the venting was denied by the U.S. Court of Appeals on June 26, and the venting began two days later. In November 1980, the Court of Appeals issued the "Sholly decision," which required the NRC to hold hearings before issuing a license amendment such as the one that permitted the containment venting. An appeal was filed on this decision with the U.S. Supreme Court, thus staying it. The Supreme Court vacated the Sholly decision in 1983 after a new law went into effect that rendered the decision moot. Against this backdrop, the TMI cleanup began in earnest.

A critical question was how to pay for what would be a major effort stretching over ten years. Only weeks after the accident, the company faced bankruptcy. The cost of replacing the electricity from TMI's units ran to $700,000 per day. Insurance would cover $300 million of what was expected to be a $1 billion price tag. Metropolitan Edison, the operator, was denied emergency rate relief by the Pennsylvania Public Utility Commission in September 1980, making the need for some plan to make up the $700 million shortfall immediate and critical.

In July 1981, Governor Thornburgh proposed a plan for sharing the cost of the cleanup among several parties, including Metropolitan Edison, the Department of Energy, General Public Utilities, and the American Nuclear Insurers. Utility customers in Pennsylvania and New Jersey would also bear a portion of the cost in their power bills. The ultimate objective of the Thornburgh plan was to reduce the radiation hazard resulting from the accident to an acceptably low level. All fuel—as well as much of the contamination—was to be removed from the reactor and associated systems, where it would pose no threat to the environment or to the public.

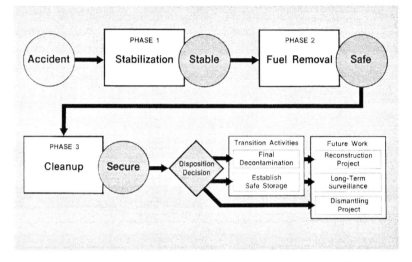

Figure 9   Recovery program. (Diagram from The Pennsylvania State University TMI-2 Recovery and Decontamination Collection, University Libraries, University Park, Pennsylvania.)

The cleanup was divided into three phases. The goal of Phase 1 was to stabilize conditions inside the reactor building sufficiently so that cleanup activities could proceed. This involved continuing to be able to control the reactor, gaining access to the inside of the containment building, beginning the initial decontamination, processing contaminated water, and storing waste. During Phase 2, workers would collect and encapsulate the fuel, removing it from the reactor and associated systems. To accomplish this, cleanup officials had to determine the state of the core, identify the precise location of all the fuel, and reduce radiation readings so that workers could safely move throughout the plant. The plant would then be ready for Phase 3—the decontamination phase.

The goals of Phase 3 were to achieve maximum decontamination and to secure systems and facilities so that they posed virtually no hazard to the public or to the environment. During this phase, the reactor would be disassembled and the plant placed in safe storage mode (called "post-defueling monitored storage"), with all remaining radioactive material under regular surveillance, until final plant disposition. To this day, TMI Unit 2 is in post-defueling monitored storage and will remain so until the site is decommissioned many years from now.

Peep Show and Quick Look

The first look inside the TMI reactor containment building—dubbed the "peep show"—occurred about nine months after the accident. While work in the auxiliary building was proceeding, no one had been able to look inside the containment area, which meant that no one knew for sure the extent of the damage. For example, had the hydrogen burn destroyed vital equipment needed for cooling the core and for defueling?

A small television camera was inserted into a cable conduit in the containment wall. Normally these conduits provide a sealed, leak-proof passage for electrical cabling running to and from the reactor building. A spare conduit was removed, and the TV camera was inserted into it and then through to the containment area.

What they saw surprised everyone. Most startling was the fact that it was raining inside the building. The humidity was high because of the large pool of water in the basement. The height of the building and the location of air coolers, which were still operating inside the containment, combined to create an unusual microclimate—and gave the reactor the appearance of a metallic rainforest. Otherwise, the containment area was remarkably clean, almost as if it were brand new. The same space that had been filled with radioactive gas and burning hydrogen was now shiny.

Once visual contact had been established, the next objective was to stabilize the containment area. This involved a number of activities designed to regain a reasonable degree of reactor control and control of water inventory, reactor coolant system pressure and temperature, loose fission products, and radioactive waste. Most important, person-

nel needed to gain easier access to the auxiliary building, which had
been severely contaminated during the accident—so contaminated, in
fact, that entries into the building required personnel to be outfitted
fully in anticontamination clothing. "Dress out" involved the use of air
breathing packs and triple anticontamination clothing (see Plate 7).

Because dress out was time consuming, much of the early effort
was aimed at decontamination so that workers could have easier
access. In most cases this involved simply wiping down surfaces to
remove contamination. Other methods, such as water lancing, power
washing, applying a strippable coating, floor scabbling, and
steam/vacuum surface cleaning, were also employed. By November
1979, the cleanup had progressed to the point where respirators instead
of self-contained breathing apparatuses could be used. Not until a year
after the cleanup began, though, could personnel access the less contaminated areas without respirators. The principal contaminants were
iodine-131 and cesium-137 (both fission products).

A major obstacle to cleanup was the contaminated water in the auxiliary building. Hundreds of thousands of gallons of water had poured
into the basement when the pressure-relief devices on the basement
collection tank had opened. The existing radioactive waste processing
system, located mainly in TMI Unit 1 but shared by both units, was not
adequate for treating the high levels of contamination in the water, so
new water treatment systems had to be installed. A portable ion
exchanger system developed by the Epicor Company was brought in to
process the less-contaminated water. Called EPICOR I, this system used
ion exchangers and filters to trap contaminants in water. A second ion
exchange system (EPICOR II) was later brought in to process higher
levels of contamination. Both systems, like home water treatment systems, used a series of filters and resin beds to remove contaminants—
in this case, radioactive contaminants—from the water.

A third system, called SDS (for "submerged demineralizer system"),
was later installed to handle the highly contaminated water from inside
the containment area. It consisted of two subsystems. The first used
zeolite, a special type of material that acts as a molecular filter to trap
contamination. Zeolite is very effective and, unlike the more common
resin beds similar to those in home water purification systems, it can

withstand high radiation levels. The second subsystem contained more standard resin beds like those used elsewhere in the plant. This SDS was designed to process the most highly contaminated water, so it was submerged in the refueling canal, which enabled the water in the canal to shield the workers from radiation.

The other major obstacle to cleanup was the high level of krypton-85 inside the containment building. Defueling was absolutely out of the question until this gas had been removed, so the decision was made to vent the containment. As noted earlier, this decision aroused concern within the local population, but the NRC eventually approved the venting. Containment purging began on June 28, 1980, and was completed on July 11. Approximately sixty-four thousand curies of krypton-85 was vented using the stack located alongside the Unit 2 containment building. The Environmental Protection Agency (EPA), the NRC, and the Pennsylvania BRP monitored the proceedings, as did The Pennsylvania State University, Metropolitan Edison, and a local citizens' group.

The venting proceeded without incident, and no serious radiation exposures occurred. Finally, on July 23—almost sixteen months after the accident—two technicians entered the containment area for the first time. As one of the technicians began to open the airlock, his words were, "Behrle to base—The inner airlock door is opening, over. I read 400 millirem about six feet inside the building, head height, over. I have entered the building, over." And so began the first of thousands of entries into the containment building (Holton, Negin, and Owrutsky 1990, J-1).

The cleanup project involved a number of distinct stages. Once containment venting was completed, a series of entries was made to obtain information to support the initial cleanup effort of decontamination, fuel removal, and plant requalification. This happened largely in 1980 and 1981. In 1982, a more in-depth data-gathering effort—the TMI Information and Examination Program—began. Funded jointly by General Public Utilities (GPU), the Department of Energy (DOE), the NRC, and the Electric Power Research Institute (EPRI), the effort was designed to obtain information about the accident and its consequences. The resulting data would prove invaluable for all aspects of the cleanup, including personnel protection, defueling, decontamina-tion, and waste management, not to mention the ongoing efforts to ensure that no radiation escaped into the environment.

Once the containment building had been stabilized and the prelimi-nary cleanup and dose reduction efforts were completed, cleanup offi-cials determined that they needed a better look inside the reactor vessel itself. This was to be done without removing the reactor vessel head— the large circular cover that is bolted to the vessel during normal operation. As noted earlier, the head is removed during refueling by unbolting it and lifting it off the vessel using a large crane (the polar crane that resides permanently in the containment building). Officials were unsure of the crane's condition after the accident, though, so instead of attempting to lift off the vessel head, they decided to remove one of the housings for the control rod drive mechanism and insert a miniature television camera into the reactor vessel through the hous-ing. The "Quick Look" inspection, as it came to be known, would pro-vide the very first view of the reactor since the accident and would bypass the time-consuming efforts needed to support and perform ves-sel head lift.

Opinions varied as to how the TMI core would look. Some experts predicted that the damage, while serious, would not constitute com-plete core failure, and that the camera would see a largely intact reac-tor. Others suggested that the core would be virtually unrecognizable and appear as a molten mass.

The Quick Look examinations were conducted on July 21, August 6, and August 12, 1982. As the first camera was lowered into the nor-mal core region, water conditions and the style of camera lens made it impossible for operators to see more than 3 inches (7.6 centimeters). The view gradually unfolded as the camera descended deeper and deeper and an operator called out the depth of penetration: "One foot into the core. Both cables going in. We are now two feet into the core." Yet nothing could be seen, no fuel rods, no debris, nothing. He contin-ued lowering the camera. "We are now approaching three feet. We are approaching four feet. We are now approaching five feet." Finally, the operator called out, "Got something" (General Public Utilities Nuclear Corporation 1982).

At a depth of 5 feet (1.3 meters) below the top of where the fuel should have been, the camera revealed distinguishable parts of the reactor sitting on a coarse bed of gravel and debris. A third of the core had been damaged and a large void had formed that was filled only by coolant water. What had happened to the fuel was not clear and would not be known fully for several more years, but one thing was perfectly obvious: the accident had been severe and the core damaged beyond repair. It took nearly a year, using photographs and sonar, to map the void.

The Quick Look inspections demonstrated that valuable knowledge could be obtained by inserting television cameras into the core. As a result, more Quick Looks were planned to answer long-standing questions concerning the presence of fuel in the lower portion of the reactor vessel. Normally this portion of the vessel contains no fuel, but instrumentation readings showed neutron levels well above those expected for a shut-down reactor, prompting the theory that large quantities of fuel had moved from the upper to the lower regions of the vessel. This theory met with considerable criticism at the time, but if it were true, there would be extensive damage to the entire core, not just the hotter upper regions.

It only became possible to inspect this region of the vessel in 1985, but, sure enough, the video found a large quantity of fuel in the lower vessel region. The fuel debris surrounded a nozzle that penetrates through the lower head of the vessel. The debris revealed that the fuel was once molten and must have flowed from above into the lower vessel head. In 1989, when the region was defueled, the amount of fuel in the lower head was actually learned: between nine and ten tons.

In addition to the Quick Looks, a series of core samplings and bores was performed to determine the state of the core. Much of the boring equipment was similar to that used in the commercial drilling and mining industry (see Plate 22). The core bores and subsequent video inspections revealed a region of once-molten fuel material estimated to be approximately 120 cubic feet (3.4 cubic meters) in volume extending across most of the core and varying in height from 1 to 3.9 feet (0.3 to 1.2 meters). The region was rock-like in nature and required considerable effort to drill through. The inspections showed some standing fuel at the periphery of the core region. Gone was any evidence of the control rods that provided normal control of the reactor.

The next step of the cleanup—fuel removal—could not begin until the reactor vessel was opened, permitting access to the core region. First, however, radiation levels had to be significantly reduced. During the initial entries, radiation levels of 200 to as high as 450 millirem per hour were recorded. The general philosophy was to decontaminate and/or reduce the dose and then to defuel. Efforts were first directed toward decontamination and later toward dose reduction.

The decontamination effort involved flushing the interior surfaces of the containment building, starting downward from the dome. The efforts were initially successful, but the surfaces became recontaminated by airborne radioactivity circulating through the air handling system. A series of experiments showed that it was necessary to clean the concrete surfaces and then to coat them to prevent loose contamination from becoming airborne and entering the ventilation system. Plate 10 shows two technicians preparing to pour a coating on a surface in the containment area. This proved successful in reducing the dose and avoiding surface recontamination, but not successful enough to proceed with the planned defueling. Not until early 1983 were the radiation levels lowered enough to proceed with defueling.

One of the biggest challenges to decontamination was the basement of the reactor building, which contained approximately 660,000 gallons (2.5 million liters) of highly contaminated water. In 1981, a major effort to pump out the water and decontaminate it (using the SDS system discussed earlier) began. The results were unsatisfactory, as radiation levels in the basement remained extremely high, with readings of 2 to 1,000 rem per hour. This was extremely disappointing, given the expense of the effort and the need to proceed with the defueling.

Eventually, acceptable radiation levels were achieved through decontamination efforts that included robotic debris removal. Studies of the basement sediment were done using a robot and strings of radiation detectors lowered into the basement. The robotic surveys were performed using the Remote Reconnaissance Vehicle (RRV), dubbed "Rover." Rover was a tether-controlled six-wheeled work platform (see Plate 11). It could be fitted with a variety of tools for radiation surveys, sampling, lifting, and decontamination. Using Rover, several types of decontaminations were performed, including flushing, ultra-high-pressure water scarifying of the concrete, removing sediment, and filling block walls with water and draining them to flush out contamination.

Defueling

Once radiation levels and airborne recontamination had been reduced to acceptable levels, defueling could begin in earnest. The polar crane had to be refurbished so that the head and internals of the reactor vessel could be removed. This proved to be probably the single most important event, because without the crane, removing the head would have been impossible. Repair work was no small undertaking, given the extreme environmental conditions the crane had been subjected to during the accident and the hydrogen burn. The adverse conditions included high radiation levels and continuous wetting during the time prior to containment venting and entry. In the spring of 1983, allegations of poor quality assurance caused a setback in the schedule, but the crane finally passed critical testing in early 1984. Load testing was successfully completed, allowing the lift to proceed.

In July 1984, the head was removed from the reactor vessel and placed in a shielded storage stand (see Plate 20). Care was taken to "diaper" the head so that any contamination falling off the head would remain localized. Previous examinations had shown that there was contamination on the head, so before actual defueling could begin, the internals of the reactor pressure vessel located above the reactor core (mainly the outlet plenum) had to be removed. The upper plenum was successfully removed in May 1985 and stored in a flooded portion of the refueling canal.

Defueling the reactor vessel was a long and laborious procedure, and it was carried out not by flooding the refueling canal, as is normally done for refueling an undamaged reactor, but by installing a special work platform over the reactor vessel in the refueling canal.

This approach, dubbed "dry defueling," had an advantage: the refueling canal would remain free of water, minimizing the amount of additional area exposed to contamination from the reactor vessel defueling efforts. The dry refueling canal also provided shielding from the remaining contamination in the reactor containment building and eliminated the need to decontaminate the canal once defueling was completed.

The platform was installed about 6 feet (1.8 meters) above the vessel flange, which is used to bolt the reactor head to the vessel. It could be rotated, enabling workers to access the entire vessel from the platform. From the platform they manually broke up pieces of the core and lifted them into canisters. Tools were specially developed for this work, including a customized work platform, scoops, cutters, drills, and airlift tools. Once filled, the canisters were transferred from the containment area and stored temporarily on site; they were eventually shipped to the Idaho National Engineering Laboratory.

Defueling began in December 1985 and proceeded in two phases. During the first phase, the rubble bed was removed, along with any solid items in or on the bed such as broken fuel rods, control rods, or core internals. A vacuum and a long-handled tool were used to retrieve the debris, after which it was loaded into the canisters. During the second phase, the core material below the rubble bed would be removed using long-handled tools similar to the ones used for the rubble bed. At least, that was the plan at the start of defueling.

As one might expect, not everything went as planned; there were unanticipated events that interrupted the schedule and demanded quick thinking and on-the-spot problem solving. One such unanticipated event was the unlikely discovery of microorganisms in the contaminated water in the core. These organisms had entered the reactor building from a leak in the air cooling system and were flourishing in the water despite the combination of chemicals and radioactivity in the coolant. Eventually, the water became so clouded that defueling operations had to be stopped and the water had to be treated.

The organisms consisted of a variety of types common in river water sediment, such as *Pseudomonas paucimobilis,* coliforms, and yeast. They were difficult to kill because of the chemistry needed to maintain safe conditions in the reactor vessel. Many common biocides were ineffective, and several laboratory experiments were conducted to find an effective agent. Eventually, hydrogen peroxide was used to eliminate the organisms during what was called the "bug kill."

Another setback to the defueling process occurred when workers encountered a hard crust about 6 to 12 inches (15 to 30 centimeters) below the original rubble bed. The existence of this crust was no sur-prise—it had been identified during the core characterization phase— but it was assumed that the crust was sitting on top of a void. Instead, it was a cone-shaped mass approximately 3.9 feet (1.2 meters) thick at the center and 1 to 1.9 feet (0.3 to 0.6 meters) thick at the edges. Attempts to break through the cone with available tools proved fruit-less, and it became necessary to bring in the equipment that had been used earlier to bore into the damaged core. Once broken up by boring, large, rock-sized rubble remained, which had to be further broken into smaller pieces for loading into the fuel canister.

Once the fuel was removed, the lower support structure inside the reactor vessel was cut into pieces by a plasma arc automatic cutting system and removed. This operation began in January 1988 and pro-vided access to the lower head beneath so that the next phase of defu-eling could proceed. The lower head was partially filled with debris that had fallen from the core region during earlier defueling operations as well as the original melted fuel that had flowed into the region dur-ing the accident. The loose debris was vacuumed, as it had been in the core region, and the hard debris had to be broken up for either picking up or vacuuming.

Fuel was discovered in other areas of the vessel as well, including the area behind a series of plates located outside the core, the so-called core former. In addition, the steam generators and the primary cooling system were found to contain fuel pumped there by the flowing cool-ant during the accident and by vessel defueling operations. Several methods were used to recover this fuel, including wet vacuuming, picking up the fuel manually with pincher tools, and picking it up remotely with a mini-submarine (see Plate 28).

On December 24, 1989, defueling was completed. It is thought that of 100 metric tons of fuel, 98.9% was removed, and only 1.1% (2,420 pounds, or 1,100 kilograms) was left at TMI, where it remains to this day. While 2,420 pounds may sound like a lot, it is spread over many areas, and in any one location the amount is too small to cause a nuclear accident. The fuel was shipped to the Idaho National Engineering Laboratory in shipping casks (see Plate 31), with the last shipment leaving TMI on April 15, 1990.

The only remaining question was how to dispose of the more than 2.8 million gallons (10.6 million liters) of accident-generated water. The water had been purified using either the SDS or the EPICOR system (or both), but neither system could remove tritium, a radioactive form of hydrogen. Because tritium behaves the same way, chemically, as hydrogen, no filter or purification system could remove it from the water.

What, then, to do with the water? The simplest solution would have been to dilute it and then discharge it into the Susquehanna River, but political and public opposition was so strong to this approach that it was abandoned in favor of gradual evaporation into the atmosphere. A total of 2.233 million gallons (8.4 million liters) was evaporated between January 1991 and August 1993. The remaining water is stored on site. It will be used in the future and disposed of later.

Summary

Today, Unit 2 is in post-defueling monitored storage. Efforts are under way to disassemble the plant gradually and to decontaminate the site when Unit 1 is decommissioned many years from now. This approach allows time for much of the remaining radioactive material to decay, which will simplify the eventual decommissioning.

Data gathered during the cleanup and related studies have provided an understanding of what happened in the reactor core during the accident. For the first hour and 40 minutes, the core was adequately cooled as water was pumped through it, with some of the water lost through the open relief valve. From 1 hour and 40 minutes to 2 hours and 54 minutes into the accident, the shutdown of the reactor coolant pumps caused the inventory of water to decrease as it boiled off. As the water boiled off, the reactor core was exposed, causing the fuel cladding to oxidize and the temperatures in the core to increase. Consequently, part of the core began to melt. For the next 6 minutes, the restart of one of the reactor coolant pumps forced water into the core, cooling the fuel assemblies on the edge of the core. This flow of water into the core caused the upper portion of the core to collapse and form a debris bed. The consolidated mass of core material continued to overheat, causing additional melting of the fuel.

After 3 hours and 20 minutes and up to 3 hours and 44 minutes into the accident, the emergency core cooling system operation brought water into the core. The water had the effect of cooling the upper debris bed, though it had minimal impact on the growth of the molten region. During the next 2 minutes, some of the fuel in the molten region of the core moved into the lower portion of the vessel, forming a debris bed. Between 3 hours and 46 minutes and 15 hours and 30 minutes into the accident, the core cooldown began. (A more detailed chronology is located in Appendix 1.)

The cleanup has taught us much about the accident and its aftermath. Many of its lessons have been applied to other existing nuclear facilities and those that have been constructed since TMI. One of the most frequently asked questions—one that deserves special attention— is "How close did we come to a 'China syndrome' disaster?" In this scenario, the core overheats, melts, and eventually melts the bottom of the reactor vessel and falls onto the concrete floor of the reactor building, where it burns through the concrete to the ground below. To answer this question, extensive vessel inspections were performed on the lower head of the reactor to determine its condition throughout the duration of the accident. These inspections revealed cracks in the protective metal coating of the vessel along with severely overheated regions, including one hot spot, roughly three feet (one meter) in diameter, where temperatures had reached as high as 2,012°F (1,100°C). Exactly how this hot spot formed is still unknown, but the common wisdom is that as the molten material moved from the core region to the lower head, some of it impinged on the head, causing serious overheating that in turn caused cracks in the interior coating.

So how close did we come to a "China syndrome" meltdown? No one can say for sure, but some experts think that had the accident continued for another twenty to forty-five minutes, the vessel would have heated up and the metal would have lost its strength, leading to a rupture. Rupture of the vessel would have made it impossible to refill the system with water, making it extremely difficult to cool the remains of the reactor core. Had this occurred, some believe that the core would have continued to melt, spilling molten uranium and fission products into the reactor building. Even if this had occurred, it is likely that the containment building would still have prevented large radioactive releases to the surrounding areas. Others think the presence of water in the lower part of the building underneath the vessel was sufficient to keep the vessel cool and prevent its failure.

What *is* clear is that a large portion of the core melted and flowed into the lower vessel. Most of the core debris experienced temperatures of at least 3,140°F (1,727°C), with certain parts reaching as high as 4,580–5,120°F (2,527–2,827°C). Much of the core radioactivity was released, but the conditions in the reactor prevented most of the cesium, strontium, and other materials from escaping into the containment. Only the noble gases escaped to any degree; these then flowed to the auxiliary building, where they were released into the environment. Unlike Chernobyl, which had little containment, the containment at TMI prevented most of the more serious radioactive material from threatening the population or the environment.

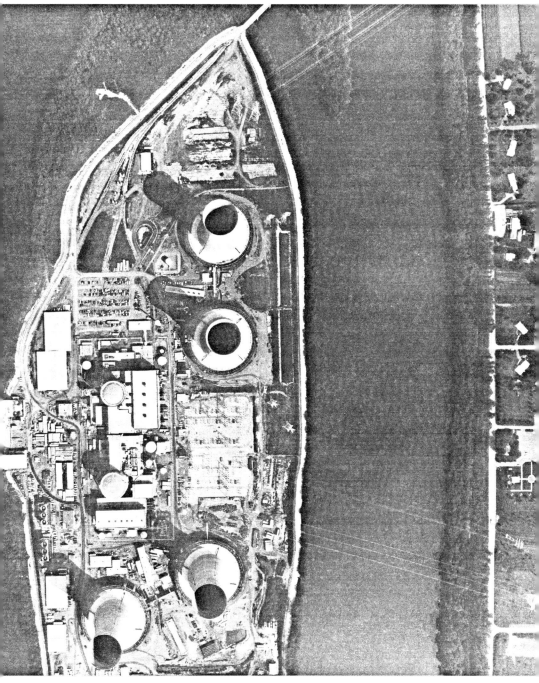

Plate 1 Aerial view of the Three Mile Island complex. TMI 2 is on the lower side of the island.

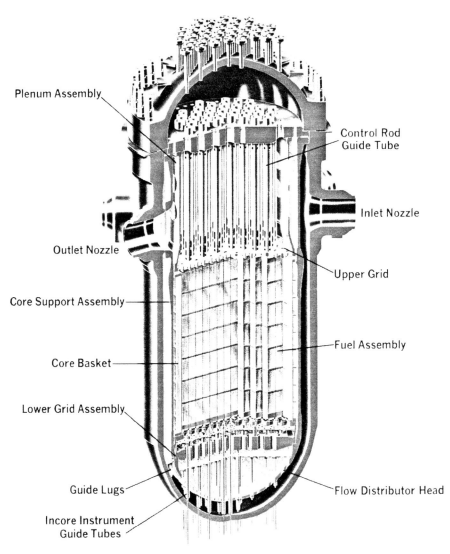

Plenum Assembly

Control Rod
Guide Tube

Inlet Nozzle

Outlet Nozzle

Upper Grid

Core Support Assembly

Fuel Assembly

Core Basket

Lower Grid Assembly

Guide Lugs

Flow Distributor Head

Incore Instrument
Guide Tubes

Plate 2 Cutaway view of a pressurized water reactor vessel similar to that used at TMI.

Plate 3 TMI 2 control room. Note the complexity of the instrumentation and controls. (Image from Rogovin et al. 1980.)

Plate 4 Inside the auxiliary fuel handling building.

Plate 5 Telephone in the containment building damaged during the hydrogen burn.

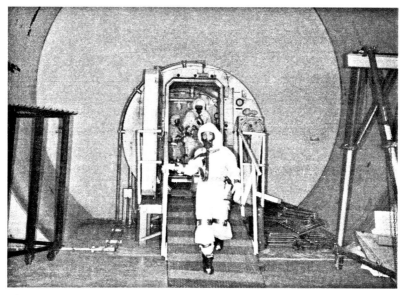

Plate 6 Technician leaving the air lock to enter the containment building.

Plate 7 Technician standing near the reactor coolant pump alignment stand inside the containment building.

Plate 8 Technicians performing decontamination work in the containment building.

Plate 9
Temporary shielding used to reduce radiation dose inside the containment building.

Plate 10 Technicians preparing for decontamination work.

Plate 11
One of the remote reconnaissance vehicles developed to
perform characterization and
decontamination work.

Plate 12 ▼
EPICOR water treatment
building and water storage
tank. (The water treatment
building is the square building with four antennas.)

Plate 13 Work being done on the in-core instrumentation and control rod electrical connections.

Plate 14
View of the rubble bed taken from the "Quick Look" TV camera inspections.

Plate 15
Damaged fuel rods and reactor components in the core.

Plate 16
Partially intact fuel rods.

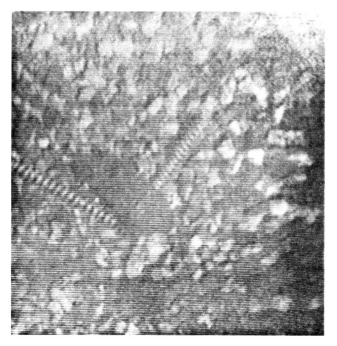

Plate 17
Rubble bed debris.

Plate 18 View of polar crane, pendant, and hook.

Plate 19 Technician examining the polar crane pendant to be used for reactor head removal.

Plate 20 The reactor vessel head after its removal, surrounded by radiation shielding.

Known Core Conditions

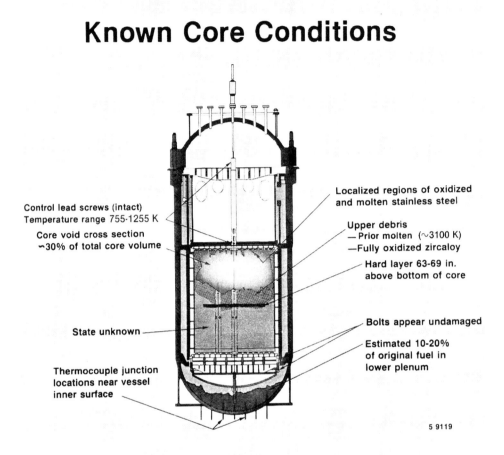

Control lead screws (intact)
Temperature range 755-1255 K

Core void cross section
~30% of total core volume

State unknown

Thermocouple junction
locations near vessel
inner surface

Localized regions of oxidized
and molten stainless steel

Upper debris
— Prior molten (~3100 K)
—Fully oxidized zircaloy

Hard layer 63-69 in.
above bottom of core

Bolts appear undamaged

Estimated 10-20%
of original fuel in
lower plenum

5 9119

Plate 21 The estimated state of the damaged reactor core before defueling began.

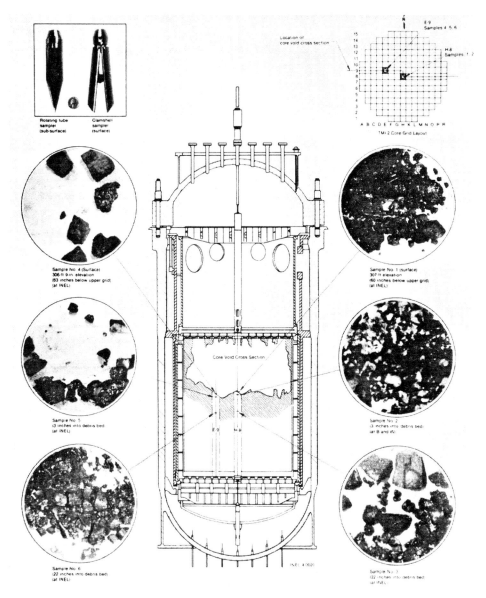

Plate 22 Views of damaged reactor core materials retrieved with a sampling tool.

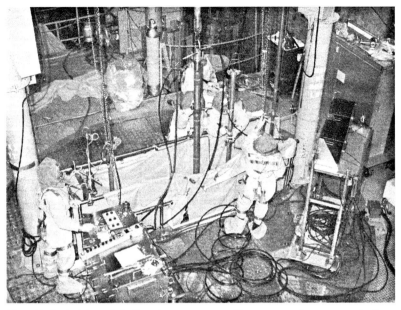

Plate 23 Technicians defueling the reactor core. They are standing on a work platform mounted above it.

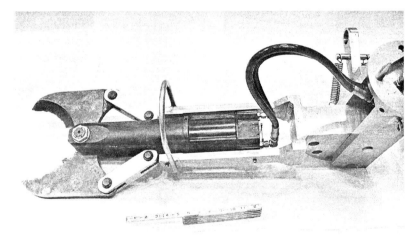

Plate 24 Cutting tool used in defueling operations.

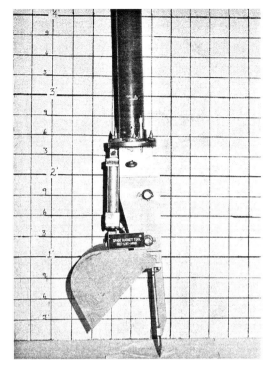

Plate 25
Spade bucket tool used in defueling operations.

Plate 26 Rubble found from melted fuel in the lower head of the reactor.

TMI-2 Core End-State Configuration

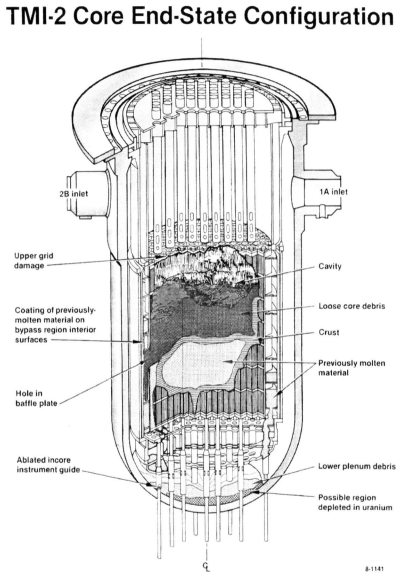

2B inlet

1A inlet

Upper grid
damage

Cavity

Coating of previously-
molten material on
bypass region interior
surfaces

Loose core debris

Crust

Previously molten
material

Hole in
baffle plate

Ablated incore
instrument guide

Lower plenum debris

Possible region
depleted in uranium

8-1141

Plate 27   The actual state of the damaged reactor core after the accident.

Plate 28 The unmanned mini-submarine used to defuel components outside the reactor vessel.

Plate 29 Crack in the inner surface of the reactor's lower head.

Plate 30 The bottom of the reactor vessel after all of the damaged fuel and internal components were removed.

Plate 31 Shipping cask used to transport damaged reactor materials from the TMI complex.

4

Media Coverage and Public Understanding

The story of the accident at TMI broke on a newscast on the Harrisburg radio station WKBO at 8:25 AM on March 28. After this short message, the news that there had been an accident at a reactor near Harrisburg, Pennsylvania, was released over the news services. Reporters from newspapers, television, and radio began to arrive to cover the story. The first reporters tended to be those who were already in the area, many of them political writers. Only later did a few reporters who were science specialists arrive.

The situation in news reporting was very different in 1979 than it is now. Looking back over twenty-five years, it is easy to forget that cable news networks did not exist: CNN was founded in 1980, a year after the accident. Most people did not have computers, never mind Internet connections. News was provided by radio, the three network news programs, local television stations, and the newspapers. Reporting often was not live but filmed earlier for evening broadcast. While reporting from the scene was commonplace, much of the reporting was based in the studio with the anchors. Reporters did not have access to computers for additional information, nor could they verify facts instantly, as they can now.

One other feature had a significant impact on the coverage. The media had covered some major stories in the twenty-five years prior to the TMI accident. The Bay of Pigs, presidential and other assassinations, riots, the fall of Saigon, the fire on the Apollo 1 launch pad, the resignation of a president: all of these were dramatic, intricate, and important stories. They were political and involved relatively well-known facts. The magnitude of the TMI accident—or at least the perceived magnitude of the accident—was beyond what the press had dealt with in the past, however. Nuclear energy was not understood by most of the reporters or by the public they would inform. Common misconceptions and fears seemed to surround this technology. After all, in the public mind, the same technology had produced the atomic bomb that leveled two cities and killed and sickened many people. Reporting the unfolding story and handling this public linkage of nuclear energy and warfare posed many difficulties for the media.

Early Nuclear Energy Coverage

The history of reporting on the atom goes back to 1939, when Thomas R. Henry attended a meeting of theoretical physicists. He heard Niels Bohr and Enrico Fermi discuss the process of fission. The reporter realized the importance of this announcement for the entire world. It was made at a meeting of specialists, however, which normally would not receive wide coverage. Thomas decided that this important story needed exposure, so he wrote an article for the *Washington Evening Star* in which he said that it was "probably the most significant event in the history of physics since the original discovery of radioactivity of uranium and radium." Thus began the coverage of modern nuclear energy (Lanouette 1989a, 7).

In the beginning, the coverage of nuclear science was positive, and there was a strong working relationship between reporters and scientists. The latter often reviewed articles before publication. Generally, the press tended to have a positive view toward the technology—until the early 1970s, that is, when the coverage tended toward the more critical and negative. Indeed, news coverage in general was more questioning of government and reported on a number of the social, political, and environmental movements of the time. Since the late 1980s, studies seem to indicate that coverage has moved toward a more balanced perspective.

One study (Media Institute 1979) indicates that there was not a great deal of coverage of nuclear energy on the national television network news programs during the 1970s. It found seven hours and fifty-two minutes of coverage between August 5, 1968, and March 27, 1979. From March 28, 1979, until April 20, 1979, during the period of time from the accident itself until media coverage lessened significantly, news programs offered five hours and thirty-nine minutes of coverage.

Most of the reporters lacked scientific and technical training, which also had an impact on coverage. As noted above, most of the original reporters at TMI were political reporters or were assigned to the story because they were already in the area. Overall, few reporters had general scientific training or any specific training in nuclear technology.

The Unfolding Story

Into this mostly rural area of Pennsylvania poured a number of reporters to cover a story that, as they all knew, had major implications. No accident of this nature had happened before. The possibility of a major disaster might exist. Local people in particular needed to be informed without undue panic but with an appropriate level of understandable information.

Many studies have looked at the coverage of TMI by the media. In addition to looking at the ways in which television, radio, and newspapers reported the accident and its aftermath, these studies have reviewed the manner in which information was shared with the media. The findings help explain some of the feelings and actions of the residents of the community and suggest changes that the media could incorporate to provide better coverage during other technological incidents.

The information sources used by the media changed during the weeks following the accident. At first, information was provided by different sources, including—at various times—Metropolitan Edison (Met Ed), the plant operator and part owner; Pennsylvania government offices; and the NRC offices. On March 31, at the request of the White House, it was decided that Harold Denton of the NRC would be the spokesperson on the accident. There were still reports coming from NRC headquarters, and later in the day, it was determined that all information would be released from Denton at the site.

One reason to centralize the source of information was to decrease the amount of conflicting data being released. Depending on which

channel or paper was consulted, the information could be very different. Paul Beers of the *Harrisburg Patriot* stated, "Had William Shake-
speare been one of the 150 reporters covering the unfolding Three
Mile Island story, he could have written, as he did in one of his plays:
'I find the people strangely fantasized; possess'd with rumors, full of
idle dreams, not knowing what they fear, but full of fear'" (Birnie
1982, 26).

This was an apt description, because there were many conflicting,
late, or false reports circulating in the press—reports that had been
delivered by credible spokespersons. Why was the information so confusing? There are several reasons, some relevant to the incident itself,
others to the organizations providing the information, and still others
related to the press.

The accident at TMI was an unprecedented technical situation.
While the industry was very committed to safety, training, and preparedness, it simply had not considered this type of accident as a likely
event. The confusion in the reactor control room as to the exact occurrences in the reactor itself became evident when reports varied and the
spokespeople could not give definitive and reassuring information. It
became even more confusing when reports emanating from different
people offered conflicting accounts. The Kemeny report notes that
there were no plans in place to disseminate information. The type of
information released—and the public's ability to understand it—would
have a significant impact on the public's response. And again, the
reporters generally had little training in technical fields. As a result,
those around the plant felt confused and alarmed. Nationwide, there
was a significant level of concern because of the confusion.

The Media Institute, a nonprofit organization that studies media
reporting, published several studies on the coverage of TMI. The data,
derived from tapes from Vanderbilt University's News Archives, indicate that there was little television coverage of nuclear energy prior to
the accident. In addition, the coverage offered very little perspective,
focusing more on the reporting of peripheral issues such as antinuclear
demonstrations or nuclear technology information without providing
context. The placement of the nuclear news during broadcasts indicated that it was not major news. The studies note that "the coverage

of nuclear energy generation has been minimal and has lacked perspective. In addition, bias often has been introduced into what should have been objective reporting, though in most cases probably unintentionally" (Media Institute 1979, 17). They conclude that "had there been more complete coverage of nuclear energy issues before March 28, 1979, the viewing public would have been in a much better position to discern the significance and potential danger from the Three Mile Island accident" (ibid., 54).

Others have looked at the television coverage of the accident and, by coding the content, have found that there were marked differences among the networks. The types of coverage were sensationalist, presenting information as emotional and of "human interest"; informative, presenting information objectively, understandably, and manageably; didactic, presenting information as educational; and "feature," presenting events within a larger picture. Dan Nimmo and James E. Combs have noted that ABC tended to be sensationalist (63% of the coverage). By comparison, NBC tended to run feature stories (54%) and didactic coverage (33%), and CBS offered more informative (44%) and feature (23%) pieces. The researchers concluded that the three networks told different stories: ABC showed the threatening view of the accident, NBC the nuclear debate, and CBS the controllability of the accident by professionals and experts (Nimmo and Combs 1982, 50).

Television played a major role in the coverage. According to polls at the time, two-thirds of the national population relied on television as a prime source of news (Media Institute 1979, 5). Radio, however, was a primary means of communication in the area around TMI during the accident. Television and radio are more timely than print, and the accident, like many similar situations, was perfect for live broadcasting. It was dramatic, with a focus on individuals and their dilemmas, and at least for television coverage, it provided distinct visual context to illustrate the situation. Philip Don Patterson has noted that there are some commonalities in disaster coverage (1987, 67–77). These include the human element, an ability to stereotype people and places, a focus on the immediate cause rather than putting the situation into perspective, the use of known experts, and the movement toward a morality story beyond the disaster itself—and TMI supplied all of these.

Newspaper coverage has also been reviewed. TMI released weekly statements to the local papers; these reports, however, were usually not used or were printed as sent. There were few stories on the reactors other than to mark significant dates in their construction or operation. One of the reasons cited for the lack of coverage was that reporters could not understand the statements generated by Met Ed. The reports were written by engineers in jargon readily understood by professionals but not by the general public. These reports did include indications of some of the problems the plant had been having, but only those conversant with the terminology could really understand what was being said. The news reporters did not pick up on the number of times there had been some type of problem with the equipment or in the operation of the reactor. The Media Institute's studies found, though, that the newspapers generally had better coverage than television because they had more time to evaluate the information and to do special reports. Moreover, they could cover more material than television could.

Indeed, a number of studies reviewed news sources and produced some interesting findings. Information was being made available by Met Ed, the state government, the NRC, and the press. Opinions vary greatly on the quality of the information provided by the different sources. While many discredited these sources, and many felt that Met Ed had lost its credibility, one study (Birnie 1982) determined that Met Ed had provided the most accurate information. State government officials presented information as it was shared with them by Met Ed and the NRC. They were responsible for using this information to make decisions that affected the population, such as the eventual partial evacuation of the area, and they were in the difficult position of having to release timely, understandable information and advice while not causing undue alarm.

One of the charges to the Kemeny Commission was to review the public's right to information. It concluded that there were major problems throughout the process and made a number of recommendations.

The Commission did not find any evidence of a cover-up of the severity of the accident, although misleading information was released at various times. It faulted the manner in which information was shared with the media, noting that too much difficult jargon had been used and that the information sources had become centralized, further limiting the media's ability to gather information. There was some sensationalizing of information, and "many of the stories were so garbled as to make them useless as a source of information" (Kemeny et al. 1979, 19).

The Kemeny report listed several findings concerning the media and the coverage of the TMI accident. The report concludes that there was no plan for providing information in the event of an accident, that official information was confused, and that the media reflected this confusion, which led to a loss of credibility for the information sources. Moreover, when the source of information to the media was centralized, the coordination was not there to ensure that this source was actually able to perform the duty adequately. The reporters were not conversant with the technology and the terminology, and they could not understand the information or present it to the public in an understandable form. The report did note, however, that the coverage generally tended to be balanced.

A great deal of interest in the role of the reporters and media was generated after TMI. Many people think about the days after the accident and remember headlines and film of people fleeing the surrounding area. With information that was conflicting and not understandable, many decided that caution dictated a fast exodus from the area. Much of this might have been averted with better information. Interestingly, additional studies looked at the accident at the Chernobyl reactor and found that some changes suggested by the Kemeny report had been implemented in the United States. Reporting on radiation was clearer and more carefully defined, graphics were informative, and terms that could be construed as alarming were not commonly used. There were additional problems with the coverage of this accident, as information was shared grudgingly with the world, but there was some evidence of improvement in the media. Since that time, there have been no other major nuclear accidents at commercial nuclear power plants; still, the trend in technical and scientific coverage has been to have experts available to explain the technology and the incidents in ways that the public can readily understand. Examples include extreme weather conditions, shuttle accidents, and the Iraq war. While the media still need improvement in some areas, the coverage of science and technology–related events has advanced since TMI caught the media relatively unprepared.

Protest Groups

Protesters against nuclear power existed long before the accident at Three Mile Island. In the 1950s, people protested in order to "ban the bomb." During the 1960s, environmental protection became a rallying point and a number of groups formed, including Friends of the Earth and Public Citizen. While nuclear energy was not their only issue, it usually was part of their agenda.

Several organizations formed around a specifically nuclear agenda. Publicizing questions about the technology, the chance of human error, and issues surrounding waste disposal, safety, and security, groups such as National Critical Mass, Mobilization for Survival, and the Clamshell Alliance took on nuclear issues. In Pennsylvania, Three Mile Island Alert was formed in 1977, two years before the accident; its Web site states that it is a "non-profit citizens' organization dedicated to the promotion of safe-energy alternatives to nuclear power and is especially critical of the Three Mile Island nuclear plant." Several other groups formed after the accident, including PANE, Newberry Township Steering Committee, and the Susquehanna Valley Alliance.

Over the years, these groups have participated in a number of activities, including demonstrations, sit-ins, organized letter-writing campaigns, speeches, debates, publications, and Web sites to promote their cause. The groups were very active after TMI. Antinuclear groups continue to be active, although they are not as publicized as they were after the accident. The Nuclear Free Hudson group has received recent attention for its protests against the Indian Point Nuclear Generating Station in New York. The Union of Concerned Scientists and the Nuclear Information and Resources Service are two that continue to monitor the industry. As the Union states on its Web site, it is "the leading watchdog on nuclear safety."

Technological Literacy

Technology is one of the core features of modern life, yet studies show that the general population's level of understanding of technology is relatively low. Technology can be defined as "the process by which humans modify nature to meet their needs and wants" (National Research Council 2002, 13). While many think of technology as products—things like computers and mobile telephones—it is actually both the products and the knowledge and processes to make them. Technology is closely linked to science, the study of the natural world. New knowledge in one area of the sciences or the technical fields can be utilized to provide new insight and advances in another subject. For example, a better understanding of materials can result in faster computers with more computing capacity. This, in turn, can be put to practical application in the control room of a reactor.

Advances in science and technology are rapid and can be complex. The specific knowledge is usually the domain of the scientific or engineering practitioner; a working knowledge of both science and technology, however, has practical implications for all. The term used to describe this necessary level of common scientific or technical knowledge is "literacy." Scientific literacy involves a basic understanding of the various scientific disciplines. Technological literacy can be defined as "an understanding of the nature and history of technology, a basic hands-on capability related to technology, and an ability to think critically about technological development" (National Research Council 2002, 11). Technical literacy and scientific literacy are the bases of well-informed decisions in both personal life and in political decisions, better jobs, and better quality of life. Many would argue that a number of options are simply not available to those without this type of literacy. Technology is that basic to much of modern life.

A technically literate person should be able to understand both the process and the basic concepts of engineering and engineering products as well as their role in everyday life. A technically literate person should also be aware of the risks of a technology as well as the trade-offs if a technology is not used, should understand how to obtain additional pertinent information in order to make technical decisions, should develop the technical skills necessary to everyday life, and should update these skills as the technology changes (National Research Council 2002, 17). The National Academy of Engineering sums this up succinctly: "technological literacy is more a capacity to understand the broader technological world rather than an ability to work with specific pieces of it" (ibid., 22). A technologically literate populace should be able to understand technology, adapt to changes, make rational technical decisions, and understand that technology is necessary to the advancement of society.

Using nuclear power as an example of this kind of literacy, we might say that a literate person should understand the basics of fuel for the reactor, the fission process, and the generation of electricity. The resultant waste products and the need for long-term storage are other aspects to be considered. The ability to understand the risks of a reactor and to assess these risks in comparison with other means of generating power is crucial to "nuclear energy literacy." There is a clear demarcation between the detailed knowledge needed to build and operate a plant and the general knowledge of the basics of nuclear power and its role in energy production. Without this basic knowledge, though, decisions about nuclear technology based on the "facts in context" cannot be made.

Unfortunately, studies show that technological literacy tends to be relatively low in the United States. Most of these studies have focused more on scientific literacy—or scientific and technological literacy combined—so it is difficult to find technology-centered studies. Interestingly, they also indicate that Americans are very interested in technology yet do not understand some of the basic principles. There are several reasons for this. Very few courses in school focus on technology, so there are few formal avenues of study. A study done by the International Technology Education Association looked at both concepts and practical application of technology and indicated that "most Americans have a very limited view of technology" (National Research Council 2002, 64). The majority of individuals polled for the study indicated that they believed "citizens should have input into technology-related decisions that affect them" (ibid., 65).

There are several intriguing paradoxes, then. Though time, money, and effort have been spent on improving science and technology classes in school and the media have increased their coverage of these topics, the studies have not indicated major changes in the public's technological literacy. There may be many reasons for this, but whatever they are, the result is still a citizenry that must make decisions and live with technology that is only partially understood. These choices have real implications for our lives: they affect our decisions about energy sources, vehicle designs, and where we will live and work, for example. In addition, though opinion polls indicate that the citizenry should be involved in making decisions about technology, these decisions are often made without pertinent or accurate facts.

One positive change since 1979 is that the media cover science and technology differently. Many newspapers and news programs regularly feature specialized science and technology stories. Reporters are better trained and may even have degrees in medicine, engineering, or some other field of science. The proliferation of science and technology television programs (and even entire channels devoted to these subjects) has provided new avenues for learning and keeping up-to-date with the rapid changes in science and technology. Interested citizens can thus follow news events and learn more about new technologies and scientific advances.

5

The Effect on the Local Community

What impact did the accident at Three Mile Island have on the health of local residents and on the environment? Many people were understandably concerned when, after the accident, Governor Thornburgh urged pregnant women and preschool-aged children to evacuate the area and radiation monitoring teams rushed to their communities. Many people—not only pregnant women and preschoolers—felt a sense of panic and left. To understand the actual impact of the TMI accident on human health and on the environment, it is important to understand radioactivity and our everyday exposures to radiation.

Understanding Radioactivity

As discussed in Chapter 1, certain elements (such as hydrogen) have isotopes that are not stable: they can transform suddenly into another form. This transformation is called radioactive decay—or radioactivity. When an atom decays, it emits energy and radiation. Radiation consists typically of alpha particles, beta particles, and gamma rays. Gamma rays are not particles but very high-energy electromagnetic radiation. (Light waves, microwaves, and X rays are forms of electromagnetic radiation, and they carry energy, just like gamma rays.) Alpha particles are just that: particles. They are made up of two protons and two neutrons. An alpha particle is identical to the nucleus of a helium atom, which has two protons. Beta particles are electrons that originate in the radioactive decay of the nucleus of an atom.

The energy released in radioactive decay appears mostly as the energy of the particle that is produced. In the case of alpha and beta particles, the energy takes the form of motion (that is, the motion of the particle). For gamma radiation, the energy is in the electromagnetic energy carried by the wave. When radiation strikes an atom or molecule, the energy that the radiation is carrying can be transferred to the atom or molecule, disrupting its normal structure. In the case of an atom, this energy can cause an electron to be ejected from the atom. When this happens, there are no longer equal numbers of positively charged protons and negatively charged electrons in the atom. The atom appears to other atoms as if it has a positive charge: it is now called an ion. Ions are very chemically reactive and can disrupt chemical processes that are essential to life. It is primarily by this means that radiation causes cell damage, cancers, or genetic defects.

A unique feature of a radioactive isotope is its eventual disappearance. When an atom decays, it disappears and is replaced by a different atom. The atom is thus transformed from one atom into another. In many cases, the new atom formed is not radioactive and is an entirely different element. Thus, over time, a radioactive material disappears and is replaced by a non-radioactive material. The time it takes for the radioactive material to disappear can be very short or very long. In the case of uranium, the time is measured in billions of years, but for a radioactive isotope of nitrogen—nitrogen-16—it is measured in tens of seconds.

The eventual disappearance of the radioactive atoms can be compared to the way chemical compounds behave. Once created, most chemical compounds will remain in the environment until they react with other chemicals. Lead, a toxic heavy metal known to cause brain damage, is a good example. The combustion of coal concentrates lead in the resulting ash. The lead can then find its way into the soil or groundwater. Because lead is relatively chemically inactive, it will remain there for a very long time. Animals and plants can absorb the lead, which eventually can result in lead exposure in humans.

The radioactive decay law governs the rate at which a collection of radioactive atoms decays. One implication of the decay law is that there is a unique decay time associated with each type of radioactive isotope. This time, called a half-life, is the time it takes for one half of the atoms to disappear by decay. After one half-life, only one-half of the atoms remain. After another half-life, one-quarter of the original atoms are left. After yet another half-life, only one-eighth of the atoms remain.

Radioactivity has a unique vocabulary. Rather than describing how many grams or ounces of a radioactive material are present, scientists usually use the term "activity." Activity refers to the number of atoms disintegrating per second. It is useful to speak about radioisotope activity because it can be related to the amount of radiation that is emitted. A special unit, the curie, describes disintegrations per second: one curie is equal to thirty-seven billion disintegrations per second. The radioactive decay law also tells us that the activity is directly proportional to the number of radioactive atoms present. Therefore, if we double the number of atoms, the activity doubles. A curie is actually a very large amount of a radioisotope, so other units, including the millicurie and the microcurie, are used. (The millicurie is one thousandth of a curie. A microcurie is one millionth of a curie.) These are the units used in the conventional system of measurement. In the newer SI system (Système International d'Unités), activity is measured in becquerels. One becquerel is equal to one disintegration per second. A curie, then, is equal to thirty-seven billion becquerels.

Background Radiation

Now that we see how radioactivity works, we must ask a different question: How much radiation is a human exposed to on a routine basis? Many think that the answer is "none"—but every man, woman, and child on earth is exposed to radiation each and every day. Most of it comes from the natural radioactive elements present in our environment. These sources include uranium in the ground, cosmic rays from space, certain isotopes of carbon and potassium that are inside our bodies, and some man-made sources, such as medical X rays, nuclear power, and airline travel.

The conventional unit used to describe the amount of radiation a person receives is the rem (in the SI system, the unit is called the sievert). The rem is a large dose; often, the millirem (mrem), or one-thousandth of a rem, is used instead. Figure 10 gives the percentages of dose from natural and man-made sources that we encounter in our environment.

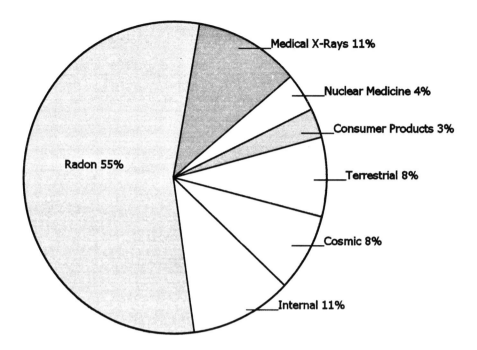

Figure 10 Sources of radiation in the environment. (Diagram from the Nuclear Regulatory Commission; see the "Glossary," s.v. "exposure," at http://www.nrc.gov/reading-rm/basic-ref/glossary/exposure.html.)

The actual dose depends on where you live. For example, the average dose in the United States, from all sources, is about 300 mrem (National Council on Radiation Protection and Measurements 1987, 149), but in other parts of the world, the dose can be as high as 500 or even 1,000 mrem. The percentages in Figure 10 are fairly typical for an average person in Pennsylvania, but they are highly variable. People living in Denver, Colorado, receive about 100 mrem more per year from cosmic radiation than Pennsylvanians do, because they are at a higher average elevation and are thus exposed to more radiation from space. For Pennsylvania, the average is approximately the same as the national average. People living in and around the Reading area of Pennsylvania, however, may also receive higher doses of radiation because there are concentrations of uranium-bearing rock in that area that produce radon, a colorless radioactive gas. For Harrisburg-area residents, the average dose from terrestrial gamma rays is about 46 mrem; from cosmic rays, 42 mrem; and from internal sources, 28 mrem. That works out to an annual dose of about 116 mrem from these natural radiation sources. They also receive a dose of about 200 mrem from radon and another 64 mrem from man-made sources. Therefore, for these residents, the average annual dose is 380 mrem.

Another dose unit used to describe the collective exposure of a population is the person-rem. A person-rem is the unit of population dose and is equivalent to one person receiving a dose of one rem. If a population of one thousand people receives one millirem, then the population has received one person-rem. The dose to a population is useful in looking at the likely health effects that will occur in an exposed population.

One assumption often made about the relationship between dose and human health effects is that no matter how small the dose, there *is* a health effect—and that if one doubles the dose, the health effects will be twice as great. This model is called a linear hypothesis. It assumes that there is no threshold below which there are no health effects. The linear hypothesis is considered conservative and may overestimate the effects of radiation. A different assumption is that a healthy body can repair damage at low dose levels; this theory suggests that there is a threshold below which there are no health effects from radiation. While some studies suggest that this is a more reasonable approach, it has yet to be adopted.

Table 2   Common sources of background radiation, in millirem per year

Source	Millirem
Stone, brick, or concrete house	7.000
Food and water	40.000
Porcelain crowns/false teeth	0.070
Computer screen	1.000
Jet travel (per hour in the air)	0.500
CAT scan	110.000
Cosmic radiation (at an elevation of 1,000–2,000 feet)	31.000
Smoke detector	0.008

SOURCE: Data from American Nuclear Society, "Radiation Dose Chart"; see http://www.ans.org/pi/dosechart/docs/dosechart.pdf.

Health effects from radiation are usually divided into two classes: non-stochastic and stochastic. Non-stochastic effects are those we know will occur when an organism is exposed, and they usually result from relatively large exposures. For example, early radiologists would experience a reddening of the skin on their hands when their hands were exposed to high levels of X rays. Called erythema, this reddening is similar to sunburn and is produced when the skin is exposed to X rays at levels of 200–300 rem.

At doses below 25 rem, there are only subtle changes in the body, primarily in the chromosomes. At doses of 75–125 rem, people experience classic radiation sickness, including nausea, vomiting, and bleeding. It takes a dose of over 240 rem to cause death—and then only when the individual is not treated.

Stochastic effects are such health effects as cancers and mutations. These may or may not occur after exposure, but they are more likely to occur as a result of an exposure. They can occur from large or small doses.

Radiation Doses from the TMI Accident

During the 10-day period over which the accident occurred, the residents of the TMI area received a radiation dose primarily from a radioactive isotope of xenon (xenon-133) and, to a lesser extent, from krypton-85. An estimated ten million curies of xenon were released. Trace amounts of radioactive iodine (thirty curies of iodine-131 and four curies of iodine-133) were also released (Eisenbud 1989, 515–56). Both were released into the atmosphere as radioactive gases escaped from the damaged reactor.

Xenon-133 and krypton-85 are both inert gases and disperse rapidly in the atmosphere. Xenon-133 has a half-life of 5.3 days and emits gamma and beta radiation. The gamma radiation was of most concern, because it has a high energy and easily penetrates the body, giving a dose to sensitive internal organs. The beta particles are not very penetrating and thus do not pose a significant health hazard. Krypton-85 emissions have effects similar to those of beta particles.

The first dose estimate was provided in May 1979 by a group of nuclear experts, the Ad Hoc Population Dose Assessment Group, whose members were drawn from the NRC, the Department of Health, Education, and Welfare, and the EPA. According to this estimate, the collective dose to the population living within a fifty-mile radius of TMI was 3,300 person-rem. (This figure does not include doses from the subsequent cleanup, doses that were estimated to be small.) Another estimate—of 2,000 person-rem—was provided by the DOE. Using the higher estimate, the average dose to an individual in this area was 1.5 mrem. The maximum dose to any individual was estimated to be less than 100 mrem—or about the same as the additional annual dose received by someone living in Denver instead of in Harrisburg. It should also be noted that this maximum dose was for a hypothetical individual who remained just outside the plant property and was exposed for the entire 10-day period from March 28 to April 7 (Nuclear Regulatory Commission, Ad Hoc Interagency Population Dose Assessment Group 1979, 2).

A later estimate, concentrating on the area within a five-mile radius of TMI, put the average likely whole-body dose for individuals at 9 mrem. The average maximum dose to an individual was put at 25 mrem (Gur et al. 1983). To place these doses in better perspective, recall that individuals living in the Harrisburg region can expect to receive an annual dose of 116 mrem simply from terrestrial gamma rays, cosmic rays, and internal sources.

Using the linear model mentioned earlier, the Ad Hoc group estimated that the radiation from TMI would cause approximately one additional cancer death among the 2.2 million people within a fifty-mile radius of the plant. In a population this size, we would expect to see about 540,000 cancers develop and about 325,000 cancer deaths. These projections were supported by estimates made by other expert panels. The conclusion drawn at that time was that the risk from the accident was not significant. Concern over the health of the residents continued, however, and a number of follow-up studies were performed.

After the accident at TMI, researchers gathered information on the physical and psychological health of individuals, radiation levels from various locations, and wildlife and farm data. The data were reviewed carefully using statistical analysis. While personal observation can seem to reveal what appears to be an abnormal number of occurrences, statistical analysis generally moves beyond anecdotal information and observation to a systematic study in order to determine whether there is a real, "significant" difference between observed numbers and expected numbers. Statistical significance refers to a genuine difference from the expected value, not a difference caused by random error or a minor difference that could logically occur without an external cause. In analyzing the data, the statistician devises a hypothesis concerning cause and effect and determines the variance. (Variance is a measure of dispersion—that is, the difference between the value of what is measured and the value that was expected.) If the variance is large enough, then it indicates that there may be a cause-and-effect relationship between the particular characteristics (or "variables") studied.

One basic principle is to look at a sample of an entire group (also called the "population"). In this case, the researchers reviewed data on radiation levels and various health effects. Their findings were compared to "normal," expected values to see whether there was a difference between these groups of people and, if so, whether the difference was meaningful (that is, statistically significant). The TMI cohort discussed below is a sample of the population around Three Mile Island.

The Effects

Health Studies

During the accident, area residents and officials were primarily concerned by the potential threat to public health posed by the TMI radiation releases. It thus seemed prudent to track the health of residents living near the reactor. Shortly after the accident, the Pennsylvania Department of Health conducted a massive survey of the approximately 36,000 people living within a five-mile radius of TMI: this number represented almost 94% of the total population of that area. Detailed data were gathered on medical history, previous radiation exposure, education, occupation, smoking behavior, and travel in and out of the affected area during the accident period. This collection of information is known as the TMI Population Registry, and the individuals within the group are often referred to in studies as the "TMI cohort." The plan was to update the health and mortality data periodically so that they could be used in future studies.

A number of health studies have been conducted over the years to determine the effects that the TMI accident may have had on the local population. Before examining these studies, it is important to understand the difficulties posed to the researchers. The Ad Hoc group had already predicted that there would be an extremely small increase in cancers—possibly one additional death out of 325,000 cancer deaths. The average dose received by the people living in a five-mile area was likely to be no more than that received by someone who experiences a chest X ray. Furthermore, many factors influence a person's cancer risk: lifestyle, occupational exposure, and smoking, for example. The researchers thus had to consider many factors in an effort to detect what could be an extremely small impact, if any. The other problem is the latency period for the development of cancers among exposed individuals. Typically, such cancers do not show up until twenty to thirty years after exposure.

A team of researchers from Columbia University performed one such study. The researchers' goal was to determine whether elevated stress levels in people living in the vicinity of TMI caused any change in the incidence of cancer (Hatch et al. 1991). Hospital records and death certificates were examined for 160,000 individuals living within ten miles of the plant. The study covered the years 1979 to 1985 and found an association between cancer rate and proximity to the TMI reactor (both before and after the accident). There was a noticeable increase in cancer for 1982–83 as well as a subsequent decline in 1984–85. The researchers postulated that the temporary increase was most likely due to the increased amount of health screening that individuals underwent, which resulted in the discovery of more preexisting cancers that were not attributable to the accident.

In another study, a team of epidemiologists from the University of North Carolina reanalyzed the data used from the Hatch study and issued new findings (Wing et al. 1997). The new analysis included a broader range of cancers, more socioeconomic variables, and a methodology to account for pre-accident differences in cancer cases among the geographic tracts in the study. Results showed a correlation between cancer incidence and accident dose, particularly in areas that would have been under radioactive plumes from the reactor releases. Lung cancer and leukemia incidence showed the closest association with dose.

Epidemiologists and public health specialists at the University of Pittsburgh carried out two long-term studies. The first one examined the total and cause-specific mortality experiences for a revised TMI cohort from 1979 through 1992 (Talbott et al. 2000). The initial cohort was reduced slightly—to 32,135 individuals—after various adjustments were made to the criteria. The health experiences of this group were compared to expected mortality rates among the citizens in three surrounding counties—York, Lancaster, and Dauphin. The researchers calculated standardized mortality ratios to make these comparisons. Relative risk regression was used to assess the relationship between cause-specific mortality rates and radiation exposure. The diseases of most interest were bronchial, tracheal, and lung cancers, breast cancer (female), lymphatic cancer, cancer of the central nervous system, and heart disease. Exposure to background radiation, smoking habits, and previous work in radiation-related fields were among the factors considered in the analysis.

The results showed that the TMI cohort had an elevated mortality rate compared to individuals in surrounding counties, but the difference was principally due to heart disease, a disease not associated with radiation. The mortality rate for cancers was very similar between the two groups. The study showed some correlation between lymphatic and hematopoietic tissue cancer risk in males with increasing exposure to background radiation, possibly due to indoor radon exposure. (Hematopoietic tissues are blood-forming tissues.) A trend linking increased female breast cancer risk with increased exposure to TMI radiation was noted, but for several reasons, the study did not attribute this increase to the accident. Although previous research showed a definite relationship between breast cancer and ionizing radiation, that research involved much higher doses than those recorded during the TMI accident. In addition, the researchers for the current study lacked sufficient background medical histories on the TMI cohort; such medical histories are crucial to determining risk factor adjustments. The researchers did recommend additional studies if the trend persisted over time.

The most recent University of Pittsburgh study (Talbott et al. 2003) reviewed mortality data for the TMI cohort from 1979 through 1998 and used methodology similar to the previous study. The total mortality and cause-specific mortality in the TMI cohort were compared to those of individuals residing in the surrounding three counties. The study found no significant difference in the overall cancer rates between the two groups. The TMI cohort, however, showed an elevated mortality rate that was primarily attributed to heart disease. The correlation seen in the earlier study between lymphatic and hematopoietic tissue cancers in males and background radiation was no longer present. For females, though, increased background radiation appears to be a risk factor for these types of cancers. The results also showed a slight relationship in males between increasing maximum gamma exposure and occurrences of lymphatic and hematopoietic tissue cancers. Moreover, the study indicated a trend toward increased risk of female breast cancer with increasing levels of maximum gamma exposure.

The researchers concluded that the mortality statistics provided no consistent evidence that radioactivity released during the accident had a significant impact on the mortality of the people in the cohort. They also noted that the slight trend for female breast cancer and likely gamma dose seen in their earlier study was no longer evident. They recommended continuing the collection of mortality data for the cohort and enhancing the database with additional information on personal lifestyle and medical histories. They also recommended further investigation into the relationship between background radiation and the health of the cohort as well as future analyses to look for cancers with long latency periods. The researchers believe that certain dose-response relationships cannot be definitively ruled out at this time.

Psychological Studies

Physical repercussions were not the only cause for concern in the aftermath of the accident. Researchers also examined psychological stress on the local population. Confusing, contradictory, or missing information in the days following the accident served to increase many people's psychological stress levels. A number of residents saw TMI not only as an acute crisis during the days and weeks of the actual accident but also as a long-term, chronic threat. The cleanup took years; there was public debate on the best actions to take; gases were vented; radioactive materials were removed; and there were protracted discussions concerning the eventual restart of the undamaged reactor, Unit 1. Researchers studied the residents near the reactor to determine what, if any, effects were evident from this long process.

A study by Gatchel, Schaeffer, and Baum (1985) looked at the relationship between the health of TMI residents and chronic stress. The study had a control group of thirty-one people from Frederick, Maryland, and a group of fifty people who all lived within five miles of the reactor. These groups were studied four times: before, during, immediately after, and six weeks after the venting of gases from the reactor. The subjects were tested in several different ways for a more complete picture of stress. The tests included proofreading and embedded-figure tests, self-reports on symptoms related to psychological and emotional distress, an attitude test, and a urinalysis for the compounds epinephrine and norepinephrine (biochemical indicators of stress). The results of this multifaceted study indicated that the TMI residents had stress symptoms a year after the accident. These symptoms increased before the publicized venting of the gases and returned to normal afterwards. Gatchel, Schaeffer, and Baum noted that stress—even minor levels of chronic stress, as found in the TMI residents—may have significant health effects over time.

Davidson, Fleming, and Baum's study from 1987 used the previous study's evaluation tools to look at chronic stress and sleep disturbances. The sample group consisted of thirty-five from the TMI area and twenty-one from a control group, with a two-thirds overlap between the two studies. The subjects took the same types of tests as in the earlier study. The results indicated that three years after the accident, there were still increased levels of stress and anxiety, higher levels of biochemical indicators of stress, and poorer performances on the proofreading exercise.

Dew, Bromet, and Schulberg (1987) looked at the effect of the TMI Unit 1 restart on the mental health of mothers. The study concluded that psychiatric symptoms were increased by the restart of the reactor and that symptoms after the accident were good predictors of increased stress at the restart. The authors suggested that long-term strategies using these predictors could be developed to address this pattern of increased stress levels during later times of concern.

A later study (Dew and Bromet 1993) focused on the mothers from the TMI area who had taken part in the 1987 investigation mentioned above. For this study, 267 women who had children between January 1978 and March 1979 were interviewed four times after the accident. Ten years later, 110 of these women answered a mailed questionnaire. The results of the study were consistent with other studies, which found a significant (if small) group suffering from elevated distress during the years following the accident. Several factors seemed to affect symptoms: lower education levels, previous psychiatric history, higher socioeconomic levels, and certain coping styles were all related to higher distress levels. One of the best predictors of problems over time was the behavior immediately after the accident. The researchers noted that this study provided some evidence for the design and use of interventions to address this kind of stress.

Schaeffer and Baum (1984) also investigated physiological responses to stress. Their sample of 121 subjects—38 from the TMI area and 83 from three control groups—was tested for adrenal levels. (Adrenal cortical activity is linked to stress.) It was determined that while the levels were higher in the TMI residents seventeen months after the accident than they were in the control group, the levels were still within normal ranges. The researchers cautioned, however, that even mildly elevated stress over a period of time may be cause for concern.

The common finding across these studies is that stress levels in residents of the area surrounding TMI are not significantly higher than in the control groups, but there is a *slightly* higher stress level that has remained elevated for a number of years. This symptom may affect life and health. The researchers have also noted that the long-term effects of stress may have health effects—physical, psychiatric, or emotional—that have not yet manifested themselves. It remains a topic of study and conjecture with no definitive answer.

Environmental Studies

The area around the TMI reactors is largely rural, with a significant number of agricultural businesses. Some fifty large dairy farms, plus a number of smaller dairy and crop farms, dotted the Susquehanna River Valley at the time of the accident. In addition to the farms, there were plants that processed milk for distribution (including the Hershey chocolate factory, only a short drive from the reactor). Fishing and boating are popular hobbies along the Susquehanna River, and a number of businesses cater to those visiting the area for such recreational activities. All of these businesses, of course, depend on a safe environment, so in 1979, as news of the incident spread, questions arose regarding the safety of the soil, air, and water. Farmers wondered about the dangers to their livestock, dairy products, and fields, and while people could evacuate, it was impossible to move large numbers of livestock. Some farmers did leave their farms, but the daily requirements of tending livestock made this option unthinkable for most. They stayed—but they worried about the safety of their milk and the viability of their farms. Farmers desperately needed information, as did the processing plants, the distributors, and the markets.

Studies began almost immediately on the safety of the water, food, and milk. The NRC, the Pennsylvania Department of Environmental Resources, and various DOE offices carried out monitoring. Tests were run from helicopters, and monitors were set up around the plant and in the surrounding areas. Samples of water, milk, and other foodstuffs were collected and evaluated for radioactive isotopes. Although the entire area around the reactor was monitored, special attention was paid to the path along which the plume traveled, and extra tests were taken along this path. From March 29 until April 10, 1979, 807 samples were studied in the vicinity of the reactor. These samples included rainwater, stagnant surface water, vegetation, soil, and air. Of these samples, only 27 had values greater than the lowest detectable activity level.

Several tests were immediately important to the people living near the reactor. Testing done on cow's milk indicated that the levels of iodine-131 stayed below the Food and Drug Administration's protective action levels and were considered safe. The EPA sent teams to monitor the levels of iodine and cesium. Both had low numbers, indicating the safety of the substances for consumption. For example, the general level of iodine in the milk tested was 36 picocuries per liter. The level at which the EPA normally takes action is 12,000 picocuries per liter. For cesium, the test indicated 46 picocuries per liter; the action level is 340,000 picocuries per liter. During the week of March 30, only 48 samples of milk tested positive for iodine-131. These levels peaked on March 31, April 1, and April 2, dropping in both the number of positive tests per day and in concentration levels.

Tests performed on the water of the Susquehanna and its aquatic life and vegetation yielded similar results. After April 17 (when the emergency phase of the testing ended), more than 4,000 test results were examined. They revealed levels of tritium, cobalt-60, and iodine-131 that were below EPA safety levels. In the samples of the fish and aquatic vegetation, levels indicated no increase over those expected from normal background radiation and fallout from atmospheric weapons testing. River sediments were at normal levels. Several researchers studied local wildlife as indicators of the effects of the incident. One study (Field et al. 1981) used the meadow vole, a small rodent common to the area near the reactor, as a monitoring organism. Three sites were selected: a control site 7.7 miles (12.3 kilometers) from the reactor and two test sites that were 1.4 miles (2.2 kilometers) and 1.1 miles (1.8 kilometers) from the reactor. Traps were set from April 6 to April 16, 1979, and 60 voles were caught. Tests of their thyroid glands revealed iodine-131 levels no higher than DOE estimates for the area.

A second study (Field 1993) examined white-tailed deer from different areas of Pennsylvania. From the six counties circling the reactor, 156 samples were collected; 21 samples were collected from ten counties at least 53 miles (88 kilometers) from the reactor. The tongues and mandibles were selected for examination because they are easy to obtain and are a good indicator of cesium levels. Samples collected in November–December 1979 revealed normal levels of cesium-137. In fact, levels in deer located near the reactor tended to be lower than in the control group taken from a greater distance. The composite of both groups was compared to studies of deer in other areas of the country after nuclear testing. Again, the Pennsylvania deer had significantly lower levels of cesium-137. One other study (Rogovin et al. 1980) looked at fishing in the Susquehanna River. While catch counts were low that season, there was no significant problem with the fish. One observation was that people tended to catch and release fish after the accident rather than eat their catch.

Environmental monitoring of the area surrounding TMI was not started as a result of the accident. The Code of Federal Regulations mandates monitoring the release of radioactive materials, and there was a record of the effluents dating back to the time the first reactor was brought online. This record provided a valid comparison for the post-incident readings.

Summary

Studies conducted in the aftermath of TMI have found no definitive health or environmental effects as a result of the radiation released from the reactor. Studies performed on samples of people from the local population have shown stress-induced health effects but no increased incidents of cancer. It must be pointed out, however, that many of these are longitudinal studies, which review health records over a period of time. As researchers continue to analyze findings from the TMI cohort, it is possible that certain effects may appear years or even decades from now.

6

The Impact of Three Mile Island

The late 1960s and early 1970s saw rapid expansion in the nuclear industry. More than 190 nuclear plants were ordered by the utility companies from 1965 through 1978. The utilities were confident that the large-capacity units being offered by reactor manufacturers such as Westinghouse and General Electric would be able to produce electricity at competitive rates. Public opinion was generally in favor of nuclear power expansion, at least partly due to the Arab oil embargo of 1973. The experience of sitting in lines at gas stations made people realize how dependent the United States was on foreign sources of energy. Nuclear power offered one way to start chipping away at this dependence. The fact that nuclear plants had been operating safely for a number of years also added some credibility to the industry.

The accident at Three Mile Island affected the country in a number of ways that went beyond the problem of cleaning up one damaged reactor. The accident drew attention to the nuclear industry and the way in which it had been operating. Congressional hearings, investigations by experts, and public pressure led to an extensive examination of training and operating procedures, regulations and laws, and the relationship between government agencies and the nuclear industry. All of these factors led to numerous changes in the industry and the way in which it operates—and have been responsible for improvements in its operating efficiency and safety record.

The Early Nuclear Industry

The nuclear industry had its beginnings in the United States with the passage of the Atomic Energy Act of 1946. This law transferred the development of nuclear technology from the military to civil authorities, and it established the Atomic Energy Commission (AEC) as the governing body. The AEC was charged with the dual role of continuing the development of nuclear technology to ensure the defense of the country and encouraging the peaceful use of atomic energy to improve the lives of the public. The AEC was given ownership of all nuclear materials and related technologies.

There was little interest in developing nuclear energy into a power-producing tool at that time for a number of reasons. Companies and utilities viewed nuclear as a brand-new and as yet untested technology for civilian applications. It was unclear whether nuclear would ever be developed into a competitive technology for producing electricity.

The U.S. military, however, was very interested in the new technology as a power source. The Navy sponsored a large program to develop reactors to power aircraft carriers and submarines. Budgetary constraints in the early 1950s led to the cancellation of the carrier project, but the program for the submarine reactor moved ahead. Rear Admiral Hyman Rickover, a strong proponent of the technology and its safe application, spearheaded the program. Westinghouse was the prime contractor in this effort and developed a prototype pressurized water reactor by March 1953. The USS *Nautilus,* the world's first nuclear-powered submarine, was under way on nuclear power on January 17, 1955.

A boost to the nascent nuclear industry came with the passage of the Atomic Energy Act of 1954. Under this law, private companies would more easily acquire and use nuclear materials and related technology as well as the technical information that would assist in research and development activities. The law also gave the AEC the power to create regulations and to license facilities producing and using nuclear materials. In the following year, the Power Demonstration Reactor Program was created to spur interest in nuclear energy. Under this program, the government and industry agreed to develop and test five different reactor technologies jointly. Companies were still cautious about nuclear energy at this time. The technology was new, and they had concerns over liability if an accident occurred at a nuclear plant. The Price-Anderson Act (1957) addressed this issue by limiting the liability of the nuclear industry in the event of damages resulting from an accident. Nuclear plant operators were required to obtain insurance from private companies, and the government agreed to pay all claims exceeding this amount up to a maximum of $500 million. The liability coverage limits provided by Price-Anderson have been increased over the years.

As noted earlier, the first commercial nuclear power plant came online in 1957 in Shippingport, Pennsylvania. Westinghouse designed and built the reactor using the expertise it had gained from building nuclear reactors to power Navy submarines. The Duquesne Light Company and the AEC funded the construction of the plant. Orders for new plants remained at a low level for several more years due to utility worries about the technology and the costs of constructing and operating the reactors. A breakthrough was achieved when the main manufacturers offered "turnkey" plants at a guaranteed price. This placed the entire burden of unexpected costs and overruns on the reactor manufacturers. Twelve plants were ordered under this arrangement between 1963 and 1967. The reactor vendors lost millions of dollars on the turnkey plants, but the program gave the utilities confidence and successfully stimulated the market for future nuclear plant orders.

Early Regulation

Like the industry it was charged with overseeing, the AEC's regulatory efforts developed slowly. In the mid-1950s, a utility that wanted to build a plant would work through a number of application procedures with the AEC. As a first step, the utility presented its plant design, its proposed site for the reactor, and its financing plans to the AEC. The next step involved filing a construction permit, which typically included a preliminary safety analysis report detailing the entire operations of the proposed plant. A group of experts appointed by the AEC, the Advisory Committee on Reactor Safeguards (ACRS), then reviewed the proposal and reported back to the AEC. At that point the AEC decided whether to issue the construction permit. Before a reactor could begin operations, the utility also had to apply for an operating license. This was normally done when the plans for the reactor design and operations were well established and the facility was nearing completion (Rolph 1979, 36).

The AEC's dual role of promoting nuclear energy and regulating it to provide for public safety was viewed by many as an inherent conflict of interest. Public hearings on licensing were not required or normally requested by the AEC. Procedural changes were forced upon the Commission, however, by the Price-Anderson Act. In response to this legislation, the AEC reorganized several times, making its internal divisions more autonomous, releasing more information to the public, and making public hearings part of the application process. The official public record for all the regulations and standards developed over the years by the AEC and its successors is Title 10 of the Code of Federal Regulations.

The large jump in nuclear plant orders in the late 1960s forced the AEC to expand its licensing staff to keep up with the workload. In 1965, the AEC formed an outside review panel, chaired by William Mitchell, to look into streamlining licensing procedures. The panel made a number of recommendations to the Commission on ways to improve the efficiency of these procedures: it suggested modifying the role of the ACRS and revising the licensing board hearing procedures, and it called for better coordination of reactor safety research (Walker 1992, 49).

In the late 1960s and early 1970s, various pressures were exerted on the AEC from both within and outside the nuclear industry. The number, size, and complexity of proposed nuclear reactors raised several questions over the safety of the units and the level of risk they posed to the public. Though the Commission was confident that it had sufficient technical background information to make proper licensing decisions, its critics were not. Their concerns centered on insufficient research data on loss-of-coolant accidents, the adequacy of emergency core cooling systems, and pressure vessel integrity. Environmental issues were highly visible in the United States during this period. Activists began questioning the safety of nuclear reactors and the environmental effects they might produce. These individuals and their supporters brought attention to the radioactive releases from nuclear plants and their potential health effects. Another nuclear plant by-product that came under scrutiny was the thermal pollution caused by the plants' discharge of cooling water to local bodies of water. Environmentalists argued that these warm-water discharges could alter the local ecosystems. These debates all contributed to an uneasy public perception of the nuclear industry.

The passage of the Energy Reorganization Act of 1974 was an attempt by Congress to separate the promotional aspect of the AEC from its regulatory side. The Nuclear Regulatory Commission was established to take over the regulatory duties of the AEC. The Energy Research and Development Administration (ERDA) would absorb the remaining responsibilities, which included research and development activities on nuclear weapons, nuclear reactors, and related areas. Despite this reorganization, the NRC continued to face the same criticisms and complaints that had plagued the AEC—charges of still being too closely tied to the nuclear industry and of not paying enough attention to reactor safety issues.

Investigations and Recommendations

The accident at TMI had immediate and long-lasting effects on the nuclear industry, the NRC, and the public. After the reactor had been stabilized and the threat of catastrophe dissipated, a general feeling of shock and anger remained. The public had, over the years, been continually reassured of the safety of nuclear power—and had just witnessed a life-threatening event. The scientists and engineers believed to be in control of the technology now appeared to be baffled by what had happened and why. The public favored putting everything "nuclear" on hold until the accident was understood and action was taken to prevent future incidents.

Two weeks after the accident, President Carter formed the President's Commission on the Accident at Three Mile Island and charged it with conducting a thorough investigation and assessment of the following issues:

1. the technical aspects of the accident, including potential health effects;

2. the role of the owning utility company, General Public Utilities;
3. emergency preparedness and the response of the NRC and other government offices (federal, state, and local);
4. the NRC's handling of TMI's licensing, inspections, and operations; and
5. the information delivered to the public during the event and how that might be improved.

The President's Commission was also charged with making recommendations for changes in these areas.

The Commission was chaired by John G. Kemeny, president of Dartmouth College, and it was composed of representatives from universities, private industry, organized labor, and the public. They were assisted by a knowledgeable staff and outside consultants. The Commission interviewed hundreds of individuals, held public hearings, and collected and generated a large quantity of documents during six months of work.

The findings of the group were published as the *Report of the President's Commission on the Accident at Three Mile Island—The Need for Change: The Legacy of TMI*. In the report, the Commission summarized its conclusions as follows: "To prevent nuclear accidents as serious as Three Mile Island, fundamental changes will be necessary in the organization, procedures, and practices—and above all—in the attitude of the Nuclear Regulatory Commission and, to the extent that the institutions we investigated are typical, of the nuclear industry" (Kemeny et al. 1979, 7).

The investigators found that the nuclear industry had been overly fixated on the equipment-related aspects of nuclear safety and not focused enough on human factors. They also discovered that the exceedingly complex NRC regulations did not necessarily guarantee safety. In fact, utilities found it difficult to follow and comply with the regulations. The regulations also tended to look at the worst possible types of equipment failures and to ignore the smaller types of incidents that could snowball into large problems (as had occurred at TMI).

In its review of health effects, the Commission was mainly concerned with the mental stress experienced by residents of the local area; they expected a negligible effect from the radiation releases from the accident. The emergency plans in place were judged to be inadequate, as were the communications among all levels of public officials. The handling of information was found to be lacking in many respects. Confusion and disagreement marked the official news sources, and the information given to the media was not always "translated" into understandable terminology (a situation exacerbated by the reporters' and correspondents' lack of scientific and technical knowledge). The result was a stream of often misleading and confusing television, radio, and newspaper accounts for the public.

The Commission criticized NRC operations in several areas. Licensing activities did not always pay sufficient attention to safety. Similarly, problems with operator training and errors were not given enough attention, and the NRC had no systematic way of evaluating the operating experiences of nuclear plants. The Commission's investigation of General Public Utilities' capabilities found many deficiencies in operator training and in the overall management of the plant.

The Commission's recommendations attempted to address these problems. Many of the recommendations have been accepted and adopted over the years; others have not. Broadly summarized, the Commission recommended these actions:

1. The NRC should be completely reorganized and focus its activity solely on nuclear reactor safety.

2. The nuclear industry must thoroughly change its attitude toward safety and regulation. It should institute a program to develop and apply safety standards to all aspects of nuclear plant operations. There should be increased sharing of operating experiences among plants, which would lead to improved operation and safety.

3. New operator-training institutions should be established with highly qualified instructors and high standards.

4. Emergency plans with detailed instructions to officials must be developed for all plants.

A major recommendation of the Commission that was not implemented was to abolish the NRC and to replace it with an independent agency, run by a single administrator, in the executive branch. President Carter declined to accept this recommendation and instead called for a restructuring of the NRC that included a strengthened role for the chairman.

Within several weeks of the accident, the NRC decided to conduct its own investigation. Investigators were charged with assessing the technical causes of the accident as well as the overall operation of the NRC. The law firm of Rogovin, Stern, and Huge was selected to lead the project. Staff members were chosen from a pool of experienced NRC personnel, and an additional group of twenty-one outside consultants was added. The Rogovin panel conducted hundreds of depositions and interviews, and it had access to the documents generated by other groups as well, including the President's Commission.

The principal finding of this investigation was that management problems—not equipment problems—were at the heart of nuclear reactor safety. Like the President's Commission, the Rogovin panel found the NRC incapable of managing a national reactor safety program with its existing form of governance. The panel also believed that the NRC had "virtually ignored the critical areas of operator training, human factors engineering, utility management, and technical qualifications" (Rogovin et al. 1980, 1:89).

The Rogovin final report included these recommendations:

1. The NRC must shift its resources from performing design review to monitoring operating reactors.
2. Highly qualified engineer supervisors should become part of the on-site management on all reactor operating shifts.
3. Future reactors should be sited more remotely. New emergency planning should be developed for existing reactors.
4. The licensing system should be overhauled and streamlined.
5. Quantitative risk assessment and other methods should be applied to review the safety of new reactor designs.

The report praised some of the strengths of the nuclear industry, such as the success of the reactor containment systems in protecting public health during the accident and the excellent technical support that nuclear experts around the country had contributed to the safe shutdown of TMI. The report concluded by emphasizing the immediate need for changes in the present NRC and nuclear industry. It expressed confidence that these changes would make nuclear power much safer in the future.

Political, Financial, and Legal Effects

The accident was unsettling to the public, particularly those living close to TMI. Before the accident, the American public saw a need for nuclear power but also believed that the regulations covering plant operations needed to be strengthened to improve safety. The public wanted new safety regulations even more strongly in the aftermath of the accident, though it still supported the development of nuclear power (Gallup 1980, 107). The accident was serious, but the reporting of the news media and the conflicting opinions of experts made it seem even worse. Many in the media presented overblown and exaggerated accounts of the situation, undoubtedly worsened by the sometimes incomplete information to which reporters had access. The lack of scientific training added to the problem of trying to explain highly technical information on short deadlines. People lost a certain amount of confidence in nuclear power as a safe way to generate electricity, and they became even more wary of plans to situate reactors anywhere near their homes and communities.

The attitude of the U.S. Congress—at least partly driven by the concerns of its constituents—also changed toward the nuclear industry. Since the inception of the U.S. nuclear power program, Congress had been supportive of the growing nuclear industry, though its support had been eroding for a number of years. Support was especially strong among representatives and senators with nuclear facilities in their districts and states. The Joint Committee on Atomic Energy had overseen industry affairs for years. After the accident, Congress held many hearings to get better insights into the industry, its regulations, and its future role in the economy. Today, congressional oversight of the industry is much more dispersed, with a number of committees sharing responsibilities in this area.

The accident had a major financial impact on the industry. General Public Utilities took the brunt of the hit, but the effect was widespread. After stabilizing the reactor, GPU needed to buy power from other utilities to meet the energy demands of its customers. It purchased power from utilities in the United States and Canada, and it received reduced rates from a number of these sources; it was not allowed to restart TMI Unit 1 for several years after the accident. The cost of the cleanup was staggering. By the time it was finished, in the early 1990s, approximately one billion dollars had been spent. Most of the funding was provided by GPU itself ($367 million), insurance payments ($306 million), other nuclear utilities ($171 million), the federal government ($76 million), and state taxes from New Jersey and Pennsylvania ($42 million).

The accident came at a time when the demand for new nuclear plants was already declining. The economic competitiveness of nuclear plants (compared to coal-fired plants) was slipping by the mid-1970s. The inflation rate in the United States was very high, making it more costly for the utilities to issue bonds to raise money for construction and related expenses. The utilities had contributed to the problem by overbuilding; there was not enough demand for electricity to justify the new plants. These factors led to a number of utility cancellations of nuclear plant orders before 1979, and TMI accelerated this situation. The overbuilding also affected the construction of new coal-fired power plants during this period, as several orders for new plants were cancelled. In all, seventy-one nuclear plant orders were cancelled between 1979 and 2001, and no new ones have been ordered since 1978. This has had a substantial impact on the industry.

Many plants under construction were never finished, adding up to billions of dollars in losses for the nuclear sector and associated industries. Nuclear plants that had received construction permits years earlier had their licensing delayed by a one-year moratorium. (The picture for nuclear power remained brighter in several other countries. Japan and France have aggressively expanded their nuclear power base over the years, largely due to their lack of other natural resources, such as coal and oil.)

The dearth of new plants and orders has affected not only industry but also subsidiary areas, such as higher education. For a number of reasons, interest in the study of nuclear engineering has waned since the accident. Public wariness about the field and the belief that there might not be jobs waiting for graduates have cut enrollments in these programs at universities across the country—despite the fact that there are still more than one hundred operating nuclear plants in the United States generating almost 22% of the nation's electric power.

Table 3   Reactor orders and cancellations, 1953–2001

Year	Reactors ordered	Reactors cancelled
1953–59	6	0
1960–65	14	0
1966	20	0
1967	31	0
1968	16	0
1969	7	0
1970	14	0
1971	21	0
1972	38	7
1973	41	0
1974	28	9
1975	4	13
1976	3	1
1977	4	10
1978	2	13
1979	0	6
1980	0	15
1981	0	9
1982	0	18
1983–2001	0	23

SOURCE: Data from Energy Information Administration, "Annual Energy Review 2002," Table 9; see www.eia.doe.gov/emeu/aer/txt/ptb0901.html.

The TMI accident spawned a multitude of lawsuits and legal actions. One of the largest suits (and the longest running) was filed against GPU in 1979 by people living and working within twenty-five miles of the plant. More than 2,000 individuals claimed that their health had been damaged due to the radiation released during the accident. Litigation moved back and forth between district court, circuit court, and the court of appeals, with judgments sometimes favoring the plaintiffs and, at other times, the defense. In December 2002, the U.S. Court of

Appeals for the Third Circuit refused to hear an appeal from a lower court on this case, effectively ending these lawsuits. The lower court had previously granted a summary dismissal of all claims against GPU. In another group of lawsuits, GPU and its insurers reached out-of-court settlements in 1985 with 280 area residents who had claimed injuries from the accident. The utility claimed that it was making the settlements to avoid litigation expenses and maintained that the accident did not cause permanent health effects. A portion of another settled class-action lawsuit was used to establish the Three Mile Island Public Health Fund in 1981. Proceeds from this 5-million-dollar fund were to be used for ongoing studies on the potential adverse health effects caused by the accident to residents living near TMI. A group of disgruntled shareholders sued GPU as well, claiming that the company had violated security laws by failing to disclose the risks involved in constructing nuclear power plants. The lawsuit was settled in 1983 when GPU agreed to pay 20 million dollars in stock and securities to shareholders who had invested in the company before the accident. In another case, GPU brought suit against Babcock and Wilcox, the manufacturer of the TMI reactor, asking for 4 billion dollars in damages. An out-of-court agreement between the parties was reached in 1983; under the agreement, GPU would receive 37 million dollars' worth of rebates on future services and equipment purchased from Babcock and Wilcox.

Changes in the Nuclear Industry

The entire nuclear industry was surprised by the severity of the TMI accident. Industry analyses over the years had investigated loss-of-coolant accidents and their potential consequences, but they were confident that enough safety features had been built into reactors to make these events unlikely. A combination of equipment failures, faulty operator decision making, and a poorly designed control room, however, contributed to the defeat of many of TMI's safety features. The industry realized that to prevent this from happening again and to help restore confidence in nuclear power, it had to make fundamental changes to the way the industry operated.

Arguably, the single most important action the industry took was to establish its own self-regulating body in late 1979—the Institute of Nuclear Power Operations (INPO). It was founded partly in response to the Kemeny Commission's recommendations and partly due to the realization that the most effective changes would have to come from within the industry itself. The Institute's goals were to establish industry-wide performance objectives and guidelines for nuclear power plant operations and operator training. It would also collect and analyze operating experience data and conduct evaluations of nuclear power plants. The Institute was determined to set a high standard of excellence for the industry in all aspects of its operations.

Over the years, INPO has developed into an organization respected both within the nuclear industry and by the NRC. It has set rigorous performance indicators and standards. Nuclear plants are inspected regularly and given ratings that reflect the outcome of those inspections. The ratings serve as strong incentives for the plant owners and managers to make the necessary improvements in operations. In addition, INPO has succeeded in getting plant operators to acknowledge that they have common goals and interests and that improved information sharing and cooperation are vital (Rees 1994, 176). The Institute founded the National Academy for Nuclear Training in 1985 in order to improve the training of operators throughout the industry. The Academy produces guidelines for training programs and develops criteria for the accreditation of the training programs at nuclear plants. It also provides related seminars and courses for plant managers and has instituted an outcome-based approach to training that stresses the knowledge and abilities that operators should possess. The NRC has given its approval to these training accreditation standards.

The NRC Responds

The NRC, too, developed an Action Plan in response to the recommendations made by the Kemeny and Rogovin investigations and by congressional committees (Nuclear Regulatory Commission 1980b). The Plan contained dozens of proposals to address the problem areas pointed out in the investigations. Proposals covered all operational areas—training, management, emergencies, radiation control, risk assessment, plant design, testing, inspections, and more. Quite a few of the proposals were eventually implemented by the NRC and can be seen in many aspects of industry and NRC operations. More emphasis is now placed on human factors in the design of control rooms and instruments, and operator training and staffing requirements are more stringent. Inspectors are permanently assigned to each nuclear plant to monitor daily operations and compliance. Probabilistic risk assessments are performed on each nuclear plant to identify potential vulnerabilities to severe accidents. Emergency planning has become an integral part of NRC and nuclear plant operations. Drills are conducted at each plant, in cooperation with local authorities, to improve the response to actual emergencies. The NRC also maintains an emergency office staffed twenty-four hours a day. The utilities have installed dedicated communication lines to the NRC and to other agencies in each plant. All nuclear plants are now required to have safety systems and instrumentation in place that have been tested to withstand the harsh radiological and environmental conditions that can occur at a plant during a TMI-like event. The NRC has required some refitting of existing plants with new equipment to upgrade fire protection, piping and auxiliary feedwater systems, and other safety-related features. Finally, the NRC has increased its enforcement activities, with tighter inspections and less tolerance for utility rule violations.

Summary

The U.S. nuclear power industry developed slowly during the 1950s but hit its stride in the late 1960s as the utilities gained confidence in nuclear technology. Growth in the industry stalled in the mid-1970s when nuclear plant construction costs soared, making nuclear power less economically competitive than other alternatives, such as coal. The Three Mile Island accident hit the nuclear industry hard, and it was partly to blame for the seventy-one plant order cancellations that occurred between 1979 and 2001. The accident further eroded public confidence in nuclear power and precipitated a number of high-level investigations into the industry. Although the amount of radiation exposure to the public during the incident was relatively minor, thousands of individuals sued GPU for damage to their health.

The recommendations issued by the special investigatory panels called for radical change in the industry and in the NRC. A major response from industry was the establishment of INPO, a self-regulating body. It has worked diligently to improve operator training, plant management, information sharing, and overall operating procedures. The NRC also retooled as a result of investigations and congressional hearings, and it, too, has instituted many changes affecting nuclear power and safety. Improvements have been made in operator training, control room design, emergency planning, and other areas. Inspectors from the NRC are permanently assigned to nuclear plants, and supervisors with technical degrees are now on duty during all shifts at nuclear plants.

All of these changes have contributed to higher efficiency and an improved safety record, as measured by several statistical indicators. "capacity factor" gives a relative measure of the actual power generated by a unit compared to its theoretical maximum output. For the entire U.S. nuclear industry, the capacity factor has moved from the 50–60% range in the 1970s and 1980s to 89% in 2001, demonstrating greatly increased reliability and generating efficiency. One measure of safety is the number of unplanned reactor shutdowns, or "scrams," that take place. Since 1980, the median number of scrams per nuclear plant per generating year has decreased from 7.3 to 0 in 2001 (Institute of Nuclear Power Operations 2003). (The median is the middle figure in the range of values being evaluated; half the values are above the median, and half below. The value of 0 scrams indicates that at least half of the power plants had 0 scrams during the year.) These figures seem to indicate that the industry is making progress toward safer and more cost-effective operations.

7

Energy for the Future

Energy has become a permanent part of the daily headlines in the United States. Gas prices going up or down, high or low inventories of heating oil, OPEC meetings, fuel efficiency of cars and trucks, debates over new exploration for oil, environmental problems with energy production—all of these concern our dependence on energy resources to maintain our current economy and standard of living. We all use energy resources in the home, at work, while traveling, and often while relaxing. A tremendous amount of energy is expended in agriculture and related industries to deliver the wide variety of foods that we enjoy daily to the table. Americans are the world's largest energy users: we make up less than 5% of the earth's population, but we use one-quarter of the energy consumed every year. Energy production, transportation, and distribution form a crucial part of our economy.

This chapter will look at energy use in the United States and in the rest of the world. It will review current consumption levels and projections for the future and will discuss energy resources, their availability, and their potential impact on the environment. In addition, the future of nuclear power as an energy resource will be examined in detail.

Energy Units

Energy can be defined as the capacity to do work. In the most basic terms, "work" is defined as a force applied over a distance. For example, lifting a 5-pound weight 1 foot vertically is equal to 5 foot-pounds of work. This amount of work can also be expressed in the metric system: 1 foot-pound is equal to 1.36 joules. So, 5 foot-pounds of work is the same as 6.8 joules. The British thermal unit (or Btu) is another very important unit of energy. One foot-pound is equivalent to 0.00129 Btu.

Thus, the energy expended in this example can also be expressed as 0.00645 Btu. The total energy consumption of countries is so large that the unit of choice is a quad. A quad is a quadrillion Btu, which can be written as 10^{15} Btu or 1,000,000,000,000,000 Btu.

Energy Use in the United States

In its early days as a nation, most of the United States' energy needs were met by wood. Wood was plentiful and easy to gather and use. It satisfied heating needs as well as most of the industrial applications of the time. Wood is estimated to have provided 0.249 quads of energy in 1775. The increased difficulty in acquiring sufficient wood supplies, combined with the high energy content and portability of coal, contributed to coal's rise as a favored energy resource by the late 1800s. By 1885, coal passed wood as the country's primary energy source. Coal supplied 2.84 quads of energy that year; by comparison, wood supplied 2.683 quads. The discovery of petroleum and the flexibility it offered as a fuel propelled it to preeminence in the mid-twentieth century as our main energy source. Natural gas also entered the energy mix in a significant way in the twentieth century, as did nuclear power. Several renewable sources, such as hydropower and wind power, have long histories of use in the United States, though not at the same levels as the other resources.

The total amount of energy consumed in the United States in 2001 was 97.46 quads. This huge amount of energy came from the sources listed in Table 4.

The demand for energy has risen fairly continuously over the years, and this trend is expected to continue for the foreseeable future. Between 1949 (when total energy consumption in the United States was 32 quads) through 2001, the early 1980s marked the only sustained period during which energy use *decreased*. The Arab oil embargo of 1979 disrupted oil supplies to the United States and stimulated a variety of conservation measures during this period, including improvements in automobile fuel efficiency and better insulation for buildings. The Energy Information Administration (EIA) issues an annual survey and forecast for energy every year. Its *Annual Energy Outlook* for 2003 gives projections for the growth in U.S. energy consumption through 2025. These projections will change somewhat, depending upon the price of world oil and the economic growth rate in the United States.

Table 4   2001 U.S. energy consumption, in quads

Source	Quads
Petroleum	38.46
Natural gas	23.26
Coal	22.02
Nuclear	8.03
Hydroelectric	2.38
Wood, waste, alcohol	2.87
Geothermal	0.32
Solar	0.06
Wind	0.06

SOURCE: Energy Information Administration, "Annual Energy Review 2001"; see http://www.eia.doe.gov/emeu/aer/ep/source.html.

These figures show that each energy source is expected to meet even more of our needs in the future than it does at present. Oil will maintain its primary share, because future worldwide availability is expected to be good—which will keep it competitively priced.

Table 5   Projected 2025 U.S. energy consumption, in quads

Source	Quads
Petroleum	56.56
Natural gas	35.81
Coal	29.42
Nuclear	8.43
Renewables (hydro, solar, wind, geothermal, wood, etc.)	8.26
Other	0.07

SOURCE: Energy Information Administration 2003a; available online at http://www.eia.doe.gov/oia/aeo/.

Where does all this energy get used? The largest share is expended in industry (33.7%). Transportation accounts for the next highest portion (27.5%). Residential use, which includes the heating, cooling, lighting, and electricity service in homes and apartments, uses 20.7%, and the commercial sector uses 17.9%.

The forms of energy used within these categories are quite different. As might be expected, virtually all of the energy consumed in transportation comes from petroleum products such as gasoline, jet fuel, and diesel fuel. Natural gas and electricity are the energy sources of choice at home and in commercial establishments. The bulk of the energy supplied to industry comes from natural gas and petroleum, with smaller but substantial contributions from electricity, coal, and wood.

International Energy Consumption

As noted above, the United States is the world's largest consumer of energy, but sizeable amounts are used by other countries as well. In fact, according to the EIA's *International Energy Outlook 2003,* much of the growth in energy consumption between now and 2025 is expected to occur in the developing countries of the world as their industrial, consumer, and economic capabilities expand. Energy consumption is projected to increase by 58% in this time period. World energy use was at 404 quadrillion Btu in 2001; it could reach 640 quads by 2025. Forty percent of this growth will occur in the developing areas of Asia, which is equivalent to a 3% annual growth rate.

Oil currently supplies 39% of the world's energy needs, and the total oil used in the world will grow substantially by 2025, though its contribution to the total energy picture will decrease slightly to 38%. Natural gas, which currently supplies 23% of the energy, will see a very large jump in growth that could place it at 28% by the end of this time period. Natural gas will move ahead of coal in terms of its energy contribution by 2005. Coal use will continue to grow at about 1.5% per year, but its part in world energy use will decline from 24% to about 22%.

The world's consumption of electricity generated by nuclear power will continue to increase. Nuclear plants generated approximately 353 gigawatts of power in 2001, and this level could rise to 366 gigawatts in 2025. The developing areas of the world will see the strongest growth in nuclear power, where a 4.1% annual increase is forecast.

Asia is developing nuclear capabilities at a rapid rate. Almost half of the thirty-five nuclear reactors under construction in 2003 are located in India, China, South Korea, and Taiwan. Asia's share of the world's nuclear-produced electricity will rise from 7% in 2001 to 17% by 2025. Outstripping any other nation in this time period, China's nuclear capability will expand from 2,169 megawatts to 19,593 megawatts. And during the same time period, South Korea's capacity is projected to increase from 12,990 megawatts to 27,609 megawatts; India's, from 2,503 megawatts to 6,986 megawatts. Renewable energy sources such as hydropower, wind, and solar will continue to contribute about 8% to the total energy pool through 2025.

Energy and the Environment

The energy-use figures that we have been discussing are truly prodigious quantities, and most of this energy is ultimately released through the burning of the fuel itself. The primary way to get energy out of fossil fuels such as coal, petroleum, and natural gas is through combustion, and combustion has several by-products in addition to heat. Depending upon the fuel, the combustion process can produce many compounds, including carbon dioxide, carbon monoxide, nitrogen oxide, nitrogen dioxide, and sulfur dioxide. Particles of various sizes, called particulates, are also formed during combustion. (Other energy sources can produce environmental effects as well; these will be discussed later.)

All of these gases and particulates are produced in the millions of tons per year in the United States. Their presence in the air has implications for health and the environment. Carbon monoxide, for instance, is thought to be the major cause of air pollution in cities. Automobiles produce at least half of the releases of this gas in the United States. The oxides of nitrogen contribute to the formation of smog in urban areas and are believed to contribute to the formation of ozone. The largest sources of sulfur dioxide gases are coal- and oil-burning power plants, with additional significant releases from the smelting industries. Sulfur dioxides and nitrogen oxides in the atmosphere are believed to be the primary causes of "acid rain" in various regions of the world. The presence of these compounds creates a slightly acidic rain that can affect plants and animals. Remote lakes have been particularly sensitive to the effects of acid rain; there are several hundred mountain lakes in the northeastern United States that cannot support fish or plant life. Forested regions in the eastern United States have also been damaged by acid rain.

Countries around the world release approximately 6 billion metric tons of carbon dioxide through energy use each year. There are concerns that excessive carbon dioxide emissions could lead to the eventual warming of the earth's atmosphere. Such a warming, even by a relatively small amount, could trigger adverse global climate changes. Some experts believe that the accumulation of excess carbon dioxide in the atmosphere will prevent the thermal radiation normally given off by the earth's surface from escaping into space. It is believed that gases such as carbon dioxide will readily absorb some of this infrared radiation, which will slowly contribute to the warming of the atmosphere. This process is called the "greenhouse effect," and the gases that cause it are called "greenhouse gases." If the global average temperature does rise by $3.6°F$ ($2°C$) by 2100, the major climatic effects would be a rise in global ocean levels and a significant change in rainfall patterns around the world. This could lead to changes in agricultural practices and disruptions in food production. In response to this problem, a set of recommendations calling on countries to reduce their greenhouse gas emissions was issued under the auspices of the United Nations in 1997. These recommendations are known as the Kyoto Protocol. The United States has not ratified it because of concerns that it would be too costly and that the schedule for implementation would be unattainable.

The Hydrogen Economy

Hydrogen is undergoing intense scientific scrutiny as a replacement fuel for petroleum and natural gas in the transportation sector. Hydrogen, however, is not an energy source; very little free hydrogen exists in nature. Nearly all of the hydrogen on earth is chemically bound to other elements. Water (H_2O) is a good example. To free the hydrogen, energy must be expended. The amount of energy required is very large—in many cases, equal to that released when the hydrogen is burned. Thus, hydrogen is better viewed as an energy *storage* medium. Energy would be stored by freeing the hydrogen, which would then be used as a portable fuel, replacing fossil fuels (primarily oil and natural gas) in transportation.

Currently, the main commercial source of hydrogen is methane or natural gas. The methane is cracked to release the hydrogen, which is then collected and stored for later use. The most promising use of hydrogen in transportation is in fuel cells, where a chemical reaction occurs that can produce electricity. Fuel cells have been used for many years as a portable power source. In the space program, for example, they provide power to manned spacecraft while in orbit. Nearly all the major automakers are experimenting with fuel-cell-powered vehicles. Unlike conventional batteries, fuel cells do not need to be recharged. Instead, the fuel is replenished in much the same way as automobiles are refueled with gasoline. Another advantage of fuel cells is that they do not emit greenhouse gases: their principal waste product is water. (Using the methane cracking technique to generate hydrogen does, however, produce the greenhouse gas carbon dioxide.)

One concept under investigation is the use of very high temperature nuclear reactors combined with thermochemical systems that would strip off the hydrogen from ordinary water—thus producing hydrogen and oxygen. Current reactors can be used to generate hydrogen by the electrolysis of water. Electrolysis breaks the water into its chemical components (oxygen and hydrogen). Although electrolysis produces no carbon dioxide, the hydrogen produced is not cost effective, compared with methane cracking or with conventional fossil fuels.

High temperature nuclear reactors show promise in lowering the cost of hydrogen production to close to that of methane cracking. In such a system, the reactor acts as a heat source for the thermochemical splitting process that strips off the hydrogen from water. The leading candidate thermochemical process is the sulfur-iodine process. To obtain economically competitive hydrogen production, the reactor must produce heat at temperatures of at least 1,652°F (900°C). This temperature is well beyond any current commercial reactor design. One concept for a reactor that may be able to operate safely and efficiently at these high temperatures is the very high temperature helium-cooled gas reactor. The reactor would produce helium at temperatures above 1,652°F. The helium would then be used to drive a chemical plant using the sulfur-iodine process, generating oxygen and hydrogen. (Waste heat could be used for the cogeneration of electricity, further increasing efficiency.) The hydrogen would subsequently be compressed or liquefied for use in automobiles. Either fuel cells or conventional internal combustion engines would use the hydrogen as a fuel.

The use of hydrogen in transportation would provide a high degree of energy independence, replacing foreign-produced oil as the fuel of choice for the transportation industry. There are many technological and social hurdles to the development of a hydrogen-based transportation economy, however. Tests are still needed to determine whether the sulfur-iodine process could be scaled to produce the large quantities of hydrogen needed to support the transportation industry. There is public concern over the safety of hydrogen as a fuel, too: many can recall the disaster in which the hydrogen-filled airship *Hindenburg* caught fire while landing in Lakehurst, New Jersey, killing many of its passengers. Can reactors be made that will operate safely and efficiently at these high temperatures?

To address these concerns, the DOE has undertaken a major research effort to examine and develop the technologies needed for the large-scale production, storage, transportation, and use of hydrogen. One piece of this program includes the investigation of different nuclear reactor concepts capable of producing the temperatures needed to economically produce hydrogen from water. The DOE's efforts are expected to take many years to confirm the feasibility of such systems and to identify how to distribute hydrogen to local gas stations and safely and efficiently use hydrogen in automobiles.

A large-scale switch to a hydrogen economy could have an unexpected impact on the environment, though. Hydrogen unintentionally released into the atmosphere during its production, storage, and transport could cause an increase in the amount of water vapor in the strato-sphere. This extra water vapor could cool the lower stratosphere and damage the ozone layer (Tromp et al. 2003). These effects might be mitigated somewhat if new technologies are developed that minimize the leakage of hydrogen. Soil, too, can absorb free hydrogen, and this could reduce the amount of gas reaching the stratosphere. This poten-tial problem further illustrates the necessity to continue to study new energy sources thoroughly, because every energy option comes with drawbacks as well as benefits.

Energy Prospects

We have seen what types of energy sources will be used in the next twenty years in the United States. Where will these energy supplies come from? Is the United States self-sufficient in any of these resources? Do we have to import energy? In fact, the United States *does* have sizeable reserves of certain forms of energy—but dwindling supplies of others. As a nation, the United States has been a net im-porter of energy since the late 1950s, when energy production could no longer keep pace with demand.

The mix of energy resources used in the United States has evolved over the years, constantly reacting to the availability and pricing of resources, the economy, and the international situation. This trend will continue into the future. Shortages and threats to supplies periodically spur research efforts into alternative sources and stimulate conserva-tion measures. This can be good for the overall energy market. The possibility of an energy substitute entering production can have a positive effect on the price of competing sources already in the market as the established producers seek to limit competition.

In this section, the major energy sources in the United States will be examined with regard to their current and projected roles as power resources, the availability of reserves, and the pros and cons of their production and use. Particular attention will be paid to renewable sources, because their technical basis is not as well known as that of fossil fuels.

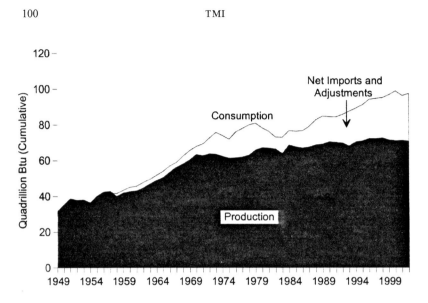

Figure 11   U.S. energy production and imports, 1949–2002. (Graph from Energy
Information Administration, "Annual Energy Review 2002"; see
http://www.eia.doe.gov/emeu/aer/pdf/pages/sec14.pdf.)

Conventional Sources

Petroleum

Petroleum (or oil) and its family of products, including gasoline, diesel
fuel, jet fuel, heating oil, and so on, represent the favorite energy
source of the United States and the world. The demand for petroleum
will continue to expand in the next couple of decades. It is an important source for the transportation sector and for industry. Growth in
demand will be especially high in transportation, which consumes two-
thirds of the petroleum products used in the country.

The United States is still a major producer of oil, but it has been
importing oil since the 1950s. U.S. production of crude oil peaked in
1970 at 9.6 million barrels per day. By 1994, the United States was
importing more than half of its petroleum supplies, and it currently
imports more than 11 million barrels of oil each day. The estimates of
oil reserves left in the United States fluctuate with exploration and the
discovery of new resources, but they are usually put at roughly a
twenty-year supply, based on expected consumption levels.

Petroleum has been a favored energy source because it can be used in all sectors of the economy, provides flexible fuels, is found in large quantities around the world, and has remained competitively priced over the years. There are several negative aspects to our use of petroleum products, though. Because needs have outstripped the domestic supply, the country is left in the vulnerable position of having to depend on other nations to keep our economy running. This has affected and will continue to affect the nation's international policies and behavior. The combustion of petroleum fuels creates gases and other by-products that can lead to local pollution problems and may contribute to atmospheric warming.

Natural Gas

Natural gas is primarily composed of methane, with lesser amounts of ethane and propane usually present. The story of natural gas production and use is similar to that of petroleum. Natural gas became an important energy source for the United States in the twentieth century and continues to be one today. Natural gases are used heavily in industry and by residential and commercial customers, and natural gas is increasingly popular as a fuel for generating electricity.

The United States was self-sufficient in natural gas production until the late 1980s. The nation currently imports more than 15% of its natural gas supply from Canada and a handful of other countries. Peak domestic production was reached in 1973, and the total production has declined slightly since then. The estimates for natural gas reserves in the United States vary widely. Conservative projections, considering proven reserves and current rates of use, predict less than ten years' worth of gas remaining. On the other end of the spectrum, more positive projections (using undiscovered resources) estimate sufficient supplies for almost fifty years.

Natural gas has become a popular energy choice because of its availability and price; it also produces smaller quantities of harmful by-products during combustion. This resource has been championed recently as a transportation fuel, and natural gas–powered vehicles are becoming a more common sight on the highway. Natural gas is a fossil fuel, however, and it shares some of the drawbacks of petroleum and coal. Potentially dangerous gaseous emissions are formed during its combustion.

Coal

Coal has been a reliable energy resource for this country since the 1800s. From 1885 through 1951, it was the nation's main source of energy. In colonial times, much of the coal that the nation used was imported. Today, however, the United States produces enormous quantities of coal and even exports some of it. The vast majority of the coal consumed in the country is burned by utilities to generate electricity. Industry uses most of the remaining portion (less than 10%).

The United States has huge reserves of coal that are projected to last for hundreds of years at current rates of consumption. Most of the coal is extracted through surface mining and the rest comes from underground mines.

The advantages to using coal as an energy source are tied to availability and price. Coal will continue to be an important component of electricity generation in coming years. It has several drawbacks, though. Underground mining is dangerous, and surface mining can disrupt local environments and scar the landscape. The burning of coal produces large quantities of carbon dioxide, sulfur dioxide, and particulate matter, which can have environmental consequences.

Renewable and Alternative Sources

A number of renewable and alternative energy sources are used in the United States. These resources currently contribute 5.7 quadrillion Btu to the nation's power supply. This figure is expected to increase to 8.3 quads by 2025. Renewable energy sources are those that can replenish themselves: they include solar, hydro, wind, ocean, geothermal, and biomass. Nonrenewables include the sources discussed previously (coal, petroleum, natural gas) and nuclear energy. Breeder reactors can generate more fuel than they use, and they can thus be considered a renewable form of nuclear power.

One category of resources, called "alternative sources," stands in contrast to some of the more conventional sources of energy, such as coal and petroleum. The major forms of alternative, non-conventional energy are non-wood biomass, solar, wind, geothermal, ocean, and fusion. Finally, there are also "green energy" resources. The major forms of green energy are windmills and solar cells. It should be noted that even green energy sources may result in some form of pollution during the manufacturing process.

Solar

Solar energy is one of the most widely known forms of alternative energy. There are two forms of solar energy—passive and active systems. The basic principles of a passive system are to utilize energy-saving features (such as building materials, siting, and design) to minimize energy use and to use energy from the sun to provide any power needed.

A passive solar system uses a solar energy collector, some means to store the energy, and a system to distribute the energy. The collector can be a greenhouse or sunspace. In some cases, the collector and the storage medium may be combined. Often, a wall, floor, water-filled structure, or other mass that can absorb the sun's rays will be used. These can also store the heat for later use. Stored heat can be released to the building by convection (the transfer of heat by motion from one substance to another) or by radiation (the movement of heat from a surface to the surrounding area).

An active system operates by the conversion of sunlight into electricity. It uses photovoltaic cells, which are usually made of silicon wafers—thin pieces of high-purity silicon. A number of these are combined into a flat structure and placed in an area that receives a significant amount of sunlight. The material is treated with impurities so that the top of the silicon wafer is negatively charged and the lower part is positively charged. This imbalance allows the flow of electrons and the generation of electricity.

Photons are packets of solar energy. When they strike a photo-voltaic cell, the photons with the right wavelength, or amount of energy, are absorbed by the solar cell. This causes the movement of electrons across the imbalanced wafer, a flow of current, and the production of electricity.

There are several advantageous features to solar energy. The solar panels are flat and relatively easy to install on a rooftop or other surface facing the sun. Equipment is minimal, because the generation of electricity is direct—no mechanical generator is needed. There are low maintenance costs and no environmental impact from the operation of the system, although manufacturing the panels results in some pollution.

While energy production is free once the system is installed, there are significant costs involved in purchasing and installing the panels and designing or retrofitting the structure for solar energy. In addition, the amount of energy that can be generated depends on both geographic location and the specific location of the structure. Some areas do not have enough sunny days to make the system practical. Other issues that must be considered are the specific siting of the structure in relation to the direction of the sun as well as trees or other structures that may block the sunlight.

Solar technology has evolved over the years: costs have come down, and efficiencies have increased. There is continued interest and research in solar both for individual homes and for large-scale power, using massive arrays. At this time, however, solar accounts for less than 1% of electrical generation.

Wind

Wind power is actually a form of solar energy. When a surface is heated unevenly by the sun, there is air movement. Water, soil, asphalt, and other surfaces each absorb solar energy differently. Warm air tends to expand and rise; cooler air moves to replace the rising warm air and results in wind. This movement can be harnessed to produce power.

Historically, windmills have been familiar structures in paintings and stories. They are associated with the Netherlands, but windmills have been used in many countries to generate power for mills and for other uses. The blades of the windmill were shaped to allow the flow of air over them, thus producing a force strong enough to turn the blades. These were connected to drive shafts, redirecting the energy to useful purposes.

Today, windmills are more accurately called wind machines. The familiar bulky structures with large blades have been replaced by streamlined structures with narrow, aerodynamically designed blades. They can be as large as a twenty-story building, with blades two hundred feet across. The more common wind machine has blades similar to those of an airplane propeller, but there are also vertical-blade machines: these have blades that connect top to bottom, bell out in the middle, and have some resemblance to the beaters on a kitchen mixer. Wind machines must be located in an area where there are no structures or natural barriers to the free flow of the wind and where the wind is both frequent enough and at a high enough speed that it can generate energy efficiently. While individuals may erect a wind machine on private property to generate power, windmill "farms" are growing in popularity in some areas. These farms have dozens to hundreds of wind machines and generate power commercially. California, Hawaii, and the Plains states utilize these farms as well as individual units.

Wind machines produce relatively little energy in the United States; they account for less than 1% of energy production. There are few environmental hazards to energy production by wind. Some problems include bird kills and the "visual pollution" that occurs when large collections of windmills are located at a single site.

Biomass

Biomass is also related to solar energy. Photosynthesis results in the growth of organic matter and the storage of chemical energy in these materials. The energy is released during burning. Biomass energy derives from manure, agricultural by-products, wood, and many waste products.

People have used biomass materials for energy throughout history. The burning of peat and wood are two ancient examples. Indeed, these continue to be burned for heat, light, and cooking. In addition, large-scale power plants have been built to generate steam from the burning of various industrial wastes. Certain industries—in particular, those using wood as a raw material—burn the wood by-products to generate power. Sugarcane producers can burn the product after the sugar is removed, and ethanol can be distilled from corn and added to gasoline to power cars.

While the production of energy from biomass materials is renewable and can address some of the issues pertaining to waste disposal, it is not necessarily an environmentally neutral technology. Air and water pollution can result. Biomass is responsible for about 3% of energy generation.

Geothermal and Ocean/Tidal Power

Geothermal energy utilizes the heat from the earth to generate electricity and to provide heat. A plant can be located close to those places where steam and hot water are trapped beneath the earth's surface and can be tapped. This kind of energy production is only feasible in a few locations—most notably in Iceland. Geothermal is an environmentally friendly technology, although there can be some small amount of local impact when the steam or hot water is removed from the ground and replaced with colder water.

Ocean power is a relatively limited alternative energy resource at this time. Two promising technologies for tapping ocean power are ocean thermal energy conversion (OTEC) and tidal power. An OTEC power plant exploits the temperature contrasts at different ocean depths to generate electricity. Tidal power relies on the movement of water to generate energy. Research is being conducted in areas where these technologies might be feasible.

Nuclear Power

Nuclear reactors have been contributing to the energy supply of the United States since the first commercial reactor came online in Shippingport in 1957. The energy supplied by nuclear plants increased substantially during the 1970s and 1980s as the construction of many plants was completed. In 2001, nuclear power generated approximately 8 quadrillion Btu, or 768 billion kilowatt hours, of power in the United States. This represented 21.5% of all the electricity generated in the nation.

For the United States, total electricity demand is expected to grow at 1.8% per year between 2001 and 2025. Much of this growth will be fueled by the increasing use of computers and electronic equipment as well as electrical appliances in residences and commercial establishments. Natural gas will increase its share of electrical generating capacity from 17% to 29% in this period, and coal's share will drop from 52% to 47%. The EIA projects that nuclear generating capacity will increase slightly during this period (from 98.2 gigawatts to 99.6 gigawatts). A decline had been previously forecast for nuclear, but increased operating efficiencies and plant licensing extensions have changed the outlook.

Nuclear Advantages

Nuclear energy offers some distinct advantages as a power source. Nuclear plants have low operating and fuel costs. The 103 plants in the United States today produce power at a very competitive rate—about 1.7 cents per kilowatt hour. The electric power production costs (per kilowatt hour) of the other major fuels are 1.8 cents for coal, 6.1 cents for natural gas, and 4.9 cents for petroleum. The nuclear costs have decreased from a high point of 3.4 cents per kilowatt hour in 1987 and reflect improved plant management and operating procedures (Nuclear Energy Institute 2003).

Uranium is the main component in nuclear fuel and is an abundant resource in the United States and several other regions of the world. With today's operating reactor technology, it is estimated that domestic uranium supplies could last for more than fifty years at the current price and rate of use. The outlook changes dramatically, however, if the nation begins to use breeder reactor technology. Breeders are designed to convert the most common form of uranium (uranium-238) into plutonium-239, a good reactor fuel. Breeders can also be designed to operate on a thorium-232/uranium-233–based fuel cycle. Thorium occurs naturally in even greater quantities than uranium. A switch to breeder reactors could theoretically extend the nuclear fuel supply for thousands of years. The low cost of uranium and concerns about the security of increased amounts of plutonium, though, have kept the interest in breeders low.

One additional advantage of nuclear power that could become very important in the future is its "cleanliness" as an energy producer. Nuclear plants do not burn fossil fuels, so they do not emit carbon dioxide, sulfur dioxide, or other gases that directly affect the environment. As nations move toward reducing their emissions of greenhouse gases, nuclear power might become an increasingly attractive energy choice. Another way of looking at this is to consider the emissions avoided by substituting nuclear plants for coal-burning ones. In 2001, by not consuming fossil fuels in power production, nuclear plants avoided the emission of 177 million tons of carbon, more than 4 million tons of sulfur dioxide, and about 2 million tons of nitrogen oxides into the atmosphere. On a global scale, the 441 operating nuclear plants prevented the emission of 600 million tons of carbon compounds.

Nuclear Disadvantages

Once they are up and running, nuclear plants can provide electricity at competitive prices. They are very expensive to design and build, however. In the deregulated electric market, utilities looking for new generating capacity are turning toward technologies that can be brought online quickly and economically. At present, natural gas and other fossil-fuel plants seem to fit these requirements best.

Nuclear plants create highly radioactive wastes in their fuel rods during power production. The spent fuel contains isotopes that will emit dangerous levels of radioactivity for years, so it needs careful handling. Used fuel rods are currently stored on site at nuclear plants. There are concerns about the safety and security of these materials; the United States is developing a long-term storage facility in Nevada for highly radioactive commercial and military nuclear wastes. It is expected that the Yucca Mountain facility will be able to isolate these radioactive wastes from the biosphere safely for years.

The fear of radiation has clouded the public's opinion of nuclear power despite the fact that the industry has a good safety record. Although the accident at Three Mile Island has apparently had negligible effects on public health, the Chernobyl accident resulted in widespread contamination and thirty-one deaths directly attributed to radiation exposure. People are concerned about a technology that has shown itself to be potentially dangerous. The public seems to agree that the nation needs nuclear power, but people would rather not live anywhere near the plants. This sentiment is so common that it has been adopted in the acronym NIMBY—"not in my backyard."

The Future of Nuclear Power in the United States

In the near term, utilities' decisions about the types of power plants to order will probably continue to be driven by purely economic considerations. Natural gas, petroleum, and coal are all predicted to remain plentiful and economical resources on the world market for the next couple of decades. Fossil-fuel plants will undoubtedly remain the most popular options for years to come.

Nuclear power may be able to capture a bigger portion of future energy production in the United States and the world if concerns over global climate change reach a critical level. There is also the possibility that unforeseen geopolitical upheavals could disrupt world energy supplies, changing the dynamics of energy availability and choice.

Future nuclear plants ordered in the United States are likely to use different designs than those currently in service. The NRC has pre-approved some new reactor designs, including the AP600, an advanced pressurized water reactor designed by Westinghouse. It has a smaller generating capacity (600 megawatts) than most operating reactors, and it relies on simplified plant systems and passive safety. Passive safety features use the forces of nature (such as gravity and natural circulation) to perform many required safety functions. The plants are projected to be easier and more economical to construct and maintain as well as safer to operate.

The pebble bed modular reactor (PBMR) is another advanced reactor concept with unique features. This helium-cooled high temperature reactor uses golf ball–sized fuel pellets. The core of the proposed reactor would contain almost a half-million fuel spheres, and helium circulating through the core would remove the heat used to generate electricity. The design is thought to be inherently safe, with no possibility of core meltdown scenarios, and the reactor can be refueled without shutting it down. The PBMR concept is being developed by a South African corporation.

The DOE is promoting research into what is called "Generation IV" nuclear plants. The work is purely conceptual at this point, but the goals are to design a plant that is highly economical and safe and that would make the diversion of materials toward weapons production very difficult. It is hoped that such a reactor would be attractive to developing nations as well as highly industrialized ones. The research effort has garnered international support.

Indeed, although the United States has the largest number of operating nuclear plants, at least thirty other nations also own reactors. Some of those nations depend more heavily on nuclear energy to generate electric power than the United States does. France has 59 reactors, and Japan, 53; these generate, respectively, 78% and 34% of their electric power needs. Both of these nations made substantial commitments to nuclear power because they have relatively small domestic energy reserves. Belgium, Bulgaria, Hungary, Lithuania, Slovakia, South Korea, and Ukraine produce at least 40% of their electric needs from nuclear energy. As mentioned earlier, total nuclear-generated electrical power will increase in the next two decades, led by growing nuclear programs in China, India, and other developing countries in Asia.

Fusion Power

The most promising form of nuclear power might turn out to be fusion energy. Fusion reactions are the processes that power the sun and other stars. The fusion reactions of interest on earth involve deuterium and tritium, isotopes of hydrogen. If a pair of deuterium and tritium nuclei or two deuterium nuclei could be made to collide at high enough energy, a nuclear reaction could take place, creating a helium nucleus and a highly energetic neutron. In a fusion reactor, the heat needed to generate electricity would primarily come from the kinetic energy of these neutrons. The roadblocks to fusion reactors are the scientific and technical challenges involved in starting and sustaining a sufficient number of these reactions in a controlled manner. A huge amount of energy must be expended to overcome the naturally repulsive force between the reacting nuclei.

For more than fifty years, worldwide research has been conducted on fusion energy. The work centers on two conceptual paths: magnetic confinement and inertial confinement. In magnetic confinement, a very high temperature plasma (ionized gas) is created and confined in the center of a vacuum chamber and heated to extremely high temperatures. Inertial confinement fusion uses a laser beam to compress a deuterium-tritium pellet to the extremely high temperatures and densities needed to ignite the fusion reactions. Both approaches have made incremental progress over the years, but much more research is needed before a fusion reactor becomes a reality.

If fusion power can be harnessed, it could satisfy energy needs well into the future, as the fuel supplies needed are widely available. Deuterium is present in seawater; tritium can be bred in reactors. Fusion reactors will be more expensive and more technically complex than fission reactors, but they will be inherently safer and produce smaller quantities of radioactive by-products.

Summary

Energy is one of the necessities of life. It is used by people across the globe for heating, cooking, transportation, commerce, agriculture, and

industry. Industrialized nations have a particularly large energy appe-
tite, and developing countries' energy consumption typically rises as
their economies grow. The world's population consumed about 404
quadrillion Btu of energy in 2001, and consumption is projected to rise
to 640 quads by 2025.

Fossil fuels such as coal, petroleum, and natural gas provide most of
the energy in the United States and other nations. This pattern is expected to continue for the next couple of decades, as worldwide supplies of fossil fuels remain plentiful. Nuclear energy and renewable
energy resources will continue to contribute to the energy mix, but on
a smaller scale than the fossil fuels.

Energy reserves are not evenly spread around the world. Indeed,
there are great disparities in reserves among nations. This fact has
allowed certain nations to become net energy exporters while others
are forced to import some or most of their energy. The United States
has historically held sizeable energy reserves, but it is also the world's
biggest consumer of energy; it has been a net energy importer since the
1950s. (The United States possesses massive reserves of coal, but
domestic oil reserves are being depleted.)

Barring unexpected geopolitical disturbances, fossil fuels may remain the most economical and flexible fuels for energy generation
during the next several decades. A major drawback of these fuels is the
copious amount of carbon dioxide and other gases that are produced
daily by their use. These compounds may be contributing to global
atmospheric warming, which could have serious consequences and
could force nations to switch to non-carbon-emitting technologies,
such as renewable resources and nuclear power.

Nuclear power produces 21.5% of the electricity used in the United
States. Nuclear plants have become very efficient in the last decade
and are now producing power at economically competitive rates. They
are expensive to build, however, and the availability of less expensive
power facilities (such as natural gas–powered plants) will probably
forestall nuclear plant orders in the near term. Long-term prospects
may be brighter for nuclear. The cost of natural gas, for instance, is
expected to rise significantly in the years ahead. Newer generations of
plants are projected to be safer and more cost-effective than current
ones, and in addition, the use of breeder reactor technologies could

extend nuclear fuel supplies for hundreds of years. Finally, the development of fusion reactors could help satisfy the world's energy needs for thousands of years.

Conclusion

If the story of Three Mile Island began with the venting of steam at 4:00 AM on March 28, 1979, it could be said to have ended with the clang of the last container filled with nuclear waste that left the reactor for Idaho in 1990, or in 1993, when the last of the water from the reactor was evaporated. Parts of the story did end on those days, certainly; many other chapters still remain to be written, however. The events of March 28 shaped lives, policies, and decisions and continue to resonate twenty-five years later.

With the benefit of this quarter-century of perspective, it is possible to revisit the accident and see it through fresh eyes. March 28 rapidly evolved into a day of terror, disbelief, and confusion. Hindsight now reassures us that the events were not catastrophic. Studies investigating physical, psychological, and environmental risks, reviewing media coverage, retracing the minute steps in the accident, and examining the reactions of the technicians and engineers proliferated in the years after the event. Health studies continue to this day. It would be fair to say that there was legitimate concern that followed the accident, when so much was unknown. Time and study, however, have substantiated the statements that the situation was not as dire as originally thought and that the industry, government, and media have learned from this accident and have taken steps to prevent another situation like TMI. A number of positive actions followed the studies of the accident. Technicians receive better training, information in control rooms is better presented, the media are better prepared to report on technological problems, and the populace has the opportunity to learn more about the science and technology that are integral to daily life.

There is no doubt that TMI was a major event in American history—and, to some extent, an important event in world history. It was a turning point for the American nuclear industry. A temporary stay was put on the licensing of reactors, and the approval process slowed down. Reactor orders were cancelled (although overbuilding capacity for nuclear plants was the primary reason for these cancellations). The undamaged and uninvolved Unit 1 reactor at TMI was not restarted until 1985, and then only after a good amount of local protest.

Throughout history, there have been comparatively few setbacks in American scientific and technical endeavors. Americans in the 1970s had become accustomed to success. Medical advances were conquering disease and life spans were increasing. After initially trailing in the space race—and the terrible Apollo 1 accident on the launchpad—there were a number of stunning successes for the United States. Americans had walked on the moon several times. Interplanetary probes were exploring the solar system and beyond. Even the potentially disastrous Apollo 13 returned home, crippled and mission aborted, but with its crew safe due to ingenuity and good science and engineering.

The events at TMI sparked many questions about the quality of our science and engineering, but when the situation was over, science and engineering had met the challenge. As studies were released over the years, it became clear that safety had been maintained by the design of the reactors and the work of the technicians and engineers. Health and safety had been protected both at the time of the accident and over the course of the cleanup after the accident.

TMI stretched the response to a potential technological problem to new limits. The scenario of the accident was one that no one thought could happen. The training and manuals did not cover this series of events, nor did they explain the procedures needed to understand the monitors and address the problems. Even with these limitations, however, radioactivity was limited to low levels. It took days to understand what was happening, but this was accomplished without danger to the residents of the area. The technology was flawed and the technicians hard-pressed, but the containment held and the engineering met the challenge. Technology did prevail over what had been almost unthinkable.

Most people learned about and followed the accident on television, on radio, or in newspapers. Most viewers and reporters were faced with a glaring fact: they knew little about this technology, which was becoming more commonplace. March 28, 1979, and the days that followed reminded Americans of the crucial role that technology played in their lives and spotlighted the fact that they simply did not have the basic understanding needed to explain how the technology worked and what the risks were. The media had difficulty understanding and reporting on the accident because they, too, did not have staff with the appropriate training in the sciences and technology. Experts were interviewed, but without training, reporters could not ask correct, detailed questions or understand whether a biased viewpoint was being advanced. It became very clear that more attention needed to be paid to scientific and technological subjects, which resulted in increased programming, more articles, and the hiring of experts by many newspapers and broadcast stations.

Metropolitan Edison, the corporate operator, had to inform the media about what had gone wrong with its facility and its steps to correct the cascading problems—a difficult and embarrassing situation. It was a burden Met Ed shouldered with the government, though, until all communication was centralized with the NRC. One of the long-term effects of the accident was that the public became more aware of risk and came to understand that technological risks needed to be explained clearly and rationally. Blind trust would not be given to anyone. Citizen monitoring groups sprang up around the plant; others that had existed before the accident gained members.

Government officials, both local and national, were caught off guard by the accident and were relatively unprepared for something of this magnitude. As a result of the accident, the government and the people are now more aware of the implications of technology in their "backyards" and have addressed their concerns to varying degrees. Citizens in the Harrisburg area, the state, and the country were forced to realize that decisions made for them by government and by corporations had a real impact on their lives and on the places in which they lived. This idea was not new, but the accident dramatized the interactions and the daily implications of new technologies, industries, and their effects on the environment. A number of technical issues, including nuclear waste and biotechnology, generated citizen interest and involvement.

Nuclear power was now looked at with a slightly more jaundiced eye; it was no longer the technology that would save us from dependence on imported oil and polluting coal and would lead us to a future of cheap, abundant energy. Technology itself was viewed as a double-edged sword by more people. It could provide advances, but these came with risks that needed to be understood. The risks now were viewed as personal and local, not remote and removed—they could affect one's children, home, and future. And these risks were not necessarily secret actions perpetrated on an unsuspecting population (like the hidden environmental disaster of Love Canal, a housing development built on dangerous toxic wastes). The risks could be a part of the routine operation of business, like reactors or genetically engineered food or antibiotics in cattle feed. The advances of technology needed to be weighed against the costs of those technologies.

That said, the events at Three Mile Island—with the benefit of historical perspective—reveal that safeguards, training, cooperation, and communication work. The immediate situation was brought under control with little real hazard, and the long-term cleanup process was done systematically, safely, and in a way that was open to public debate and scrutiny. This becomes even more evident when compared to the Chernobyl accident in 1986, which indicated a lack of communication, of safeguards at the plant, and of safety features to protect the community.

The saga of TMI continues. While studies reveal that there was little environmental damage, long-term studies of health effects are ongoing. Some possible repercussions may not appear for years, so even the twenty-year report indicates that the TMI cohort will continue to be monitored over time. The findings so far are encouraging, though, showing little medical effect on the population and relatively little effect on the psychological health of those in the studies.

In addition to the health studies, TMI will continue to have an impact on the training of people who must deal with dangerous technologies. Those involved in the nuclear industry study the scenario carefully and review the decontamination and recovery process. Technicians receive training that has been shaped by the accident and its aftermath. Expecting and being prepared for the unexpected—and even the "impossible"—are now essential skills. Risk management and the ability to convey risks to the general population rationally and understandably play an integral role in the implementation of new technologies.

Twenty-five years have passed, and an entire generation may know little about the nerve-wracking days following the accident or the years of cleanup. This is partly because the accident was contained, controlled, and cleaned up with so little long-term effect. Yet the interest in TMI continues. Of course, many people still live within sight of the four cooling towers, only two of which now spew plumes of vapor into the air. TMI continues to be a part of the residents' history and a part of everyday life.

After the accident and the cleanup process, a significant collection of videotapes, photographs, and reports documented the complicated process of examining the reactor. These were donated to the University Libraries at The Pennsylvania State University in University Park, Pennsylvania. The collection was organized and cataloged into a database available on the World Wide Web; see http://www.libraries.psu.edu/ TMI/index.htm. The TMI Recovery and Decontamination Database has a short explanation of the collection, a video database, a report database, a link by which viewers can submit questions, and a twenty-four-minute video that summarizes the decontamination project. Use of the collection has been constant since it went public in 1995, with requests for materials and questions coming from professionals in the nuclear industry, citizen advocates, and schoolchildren. Questions range from the very technical to the philosophical. Requests have come internationally. While no statistical record has been kept, the questions continue to arrive—well more than two decades after the accident.

The far-reaching legacy of TMI can be seen in the changes in the training of nuclear technicians and in the detailed, longitudinal health studies begun after the accident. It has had an impact on general attitudes and on specific procedures. Political, environmental, and energy decisions have been made with TMI as a factor. The interpretation of that legacy will largely depend on a range of issues and positions—scientific, technical, philosophical, economic, political, and personal.

TMI has been labeled the worst nuclear power accident in the United States; until 1986, some called it the worst nuclear power accident in the world. It is a milestone in technological history. The lessons learned have helped shape our world. Objective knowledge will help provide guidance for the decisions that will need to be made as we go forward into the next quarter-century.

Appendix 1: Chronology of the TMI Accident and Cleanup

This scenario was developed as part of the TMI recovery and characterization program described in Chapter 3. It represents the best estimate of times and events as they occurred during the accident and the subsequent cleanup effort.

March 28, 1979	4:00:37 AM	Feedwater pump stops; turbine is stopped
	4:00:40–43	Pressure in the primary system reaches the point at which the pressurizer relief valve opens
	4:00:45	Reactor scrams
	4:00:50	Pressure in the primary system falls to the point at which the relief valve should close
	4:01:37–04:37	Pressurizer level rises rapidly
	4:02:22	Steam generator (secondary side) boils dry
	4:02:39	Pressure in the primary system falls to 1,600 psi; emergency core cooling system pumps start
	4:03:50	Reactor coolant drain tank relief valve opens, spilling coolant into the reactor building basement
	4:04:37	Operator reduces flow of water from emergency core cooling system
	4:05:15	Operators shut down last emergency core cooling pump

March 28, 1979 (cont.)	4:06:37	Pressurizer becomes full of water
	4:08:06	Reactor building sump pump starts pumping contaminated water from reactor building basement to auxiliary building
	4:08:37	Operators restore flow of water to steam generator (secondary side)
	4:15:37	Primary system reaches pressure at which it begins to boil
	4:33:13	Reactor core thermocouples readings off the scale, indicating high temperatures in core
	5:41:37	Operators stop remaining reactor coolant system pump, stopping coolant flow to reactor
	5:52:04	Reactor coolant system loop temperatures go off the scale, higher than 620°F (327°C)
	6:18:37	Operator isolates the open relief valve
	6:39:00	Radiation monitor readings in auxiliary and reactor buildings begin to increase
	6:47:00	Instruments in core indicate core temperatures down to four-foot level are above 700°F (371°C)
	6:54:37	Operator restarts reactor coolant pump and runs it for 19 minutes
	6:55:37	Site emergency declared
	7:20:37	Emergency core cooling pumps restart automatically; operators turn pumps off again and operate them intermittently until 8:30:37

March 28, 1979 (cont.)	7:44:37	Core relocation occurs
	8:30:37	Emergency core cooling system reactivated; flow at high rate
March 29, 1979	Noon	Lt. Governor William Scranton tours TMI
	11:00 PM	Governor Richard Thornburgh told of core damage
	11:30	Hydrogen bubble detected in reactor
March 30, 1979	11:00 AM	Release of radioactive materials; local residents told to take shelter
	12:30 PM	Evacuation advisory released for pregnant women and pre-school-aged children
	2:15	Thornburgh notified: meltdown possible
	3:30	NRC notifies press: meltdown possible
	9:00	Press conference with Harold Denton and Thornburgh
March 31, 1979	11:00 AM	Met Ed announces bubble is shrinking; Denton and Thornburgh hold press conference, announcing that explosion is not possible and that President Carter to arrive next day
April 1, 1979		President and Mrs. Carter arrive at TMI
April 2, 1979	11:15 AM	Denton announces hydrogen bubble is small
April 5, 1979		Secretary of HEW Joseph Califano announces long-term health study of plant workers and pregnant women from the TMI vicinity; EPA begins milk testing program
April 8, 1979		Antinuclear rally in Harrisburg

April 10, 1979	John Kemeny, president of Dartmouth, appointed by President Carter to head investigation committee
April 1979	Cleanup begins: decontamination of auxiliary building started
April 27, 1979	TMI reactor achieves cold shutdown
November 10, 1979	"Peep show"
November 1979	Workers routinely enter auxiliary building without self-contained breathing apparatus
June 28–July 11, 1980	Containment vented
July 23, 1980	First containment entry
October 30, 1985	Defueling officially begins
November 12, 1985	Beginning of actual fuel removal
January 1990	Defueling completed
January 1991	Evaporation of "accident water" begins
August 1993	Water evaporation completed
December 28, 1993	TMI 2 placed in monitored storage

Appendix 2: Glossary

The following definitions are based on the Nuclear Regulatory Commission's glossary, which is available online at http://www.nrc.gov/reading-rm/basic-ref/glossary.html.

activation	inducing radioactivity in a material by bombarding it with neutrons or other types of radiation
activity	the radioactive intensity of a source, expressed in the number of disintegrations or transformations per unit time
alpha particle	a positively charged particle (composed of two protons and two neutrons) that is emitted by a number of radioactive substances and is equivalent to the nucleus of a helium atom
atom	the basic unit of an element; consists of a nucleus made up of protons and neutrons, surrounded by orbiting electrons
atomic number	the number of protons in the nucleus of an atom
auxiliary feedwater	backup water supply that can be used to provide water to the steam generators of a pressurized water reactor or to the reactor in a boiling water reactor, thus providing a path for cooling the reactor
background radiation	radiation that originates from a variety of sources in the environment, including cosmic rays, naturally occurring radioactive elements contained in certain rocks and geological formations, and radioactive gases in the atmosphere
becquerel	a basic unit of radioactive decay equal to one disintegration per second
beta particle	an electron or positron emitted from a substance during radioactive decay

boiling water reactor	a nuclear reactor cooled by water in which the water is allowed to boil in the core; the steam formed is used to power an electric generator
breeder	a nuclear reactor that produces more nuclear fuel than it uses
British thermal unit (Btu)	the energy required to raise the temperature of one pound of water $1°F$ at sea level
capacity factor	the ratio of the actual power generated by a plant to the plant's maximum generating capacity
cask	a heavy container, typically made with lead and steel, used to store or transport radioactive materials
chain reaction	a sequence of nuclear reactions in which neutrons produced by fission reactions produce subsequent fission reactions
charged particle	a particle whose electrical charge is not 0 (for example, an electron has a charge of -1; a proton has a charge of $+1$)
cladding	the outer metallic jacket on a nuclear fuel element
cogeneration	the simultaneous generation of electrical energy and steam energy from the same plant
cold shutdown	the state of a system, such as a nuclear reactor, when the coolant is below its boiling point
collective dose	the cumulative sum of individual doses received by a population during a period of time from exposure to a radiation source
compound	a substance consisting of different molecules that cannot be separated by physical means
condensate	the liquid produced in a condenser
condenser	a heat exchanger that converts steam back into water by removing heat
containment structure	the airtight, steel-reinforced enclosure surrounding the nuclear reactor that is designed to prevent radioactive gases from escaping into the environment during an accident

contamination	the presence of unwanted radioactive materials on surfaces, structures, or objects
control rod	a rod or tube containing a neutron-absorbing material (such as boron or hafnium) used to control the power level in a nuclear reactor
control room	the room at a nuclear plant containing most of the controls and instruments necessary to control the reactor and its related safety equipment
coolant	the material that circulates through the core of a nuclear reactor to carry away heat; coolants can be water, heavy water, air, gases, or liquid metals
cooling tower	the large concrete structures used to cool the water that condenses the steam leaving the turbines at a nuclear plant
core	the part of the nuclear reactor that contains the fuel elements, moderator, and internal support structures
core melt accident	an accident involving the melting of part of the fuel in a nuclear reactor
cosmic radiation	ionizing radiation, originating in outer space, that can consist of very high energy particles or electromagnetic energy
criticality	the condition when a reactor loses exactly the same number of neutrons that are produced, so that the power or number of fissions remains the same over time
critical mass	the minimum amount of fissionable material necessary for a self-sustaining chain reaction
cumulative dose	the total dose received by an individual from all exposures to radiation over a period of time
curie	the basic unit of radioactivity; one curie equals thirty-seven billion disintegrations per second
daughter product	the isotope produced from the radioactive decay of some other isotope

decay, radioactive   the process during which a nucleus of one type of substance transforms into another substance, accompanied by the emission of radiation

decay heat   the heat produced by the decay of radioactive substances produced in fission after a reactor has been shut down

decommissioning   the permanent closing of a nuclear plant, including the removal of all structures and the cleaning of the site to permit free public access and use of the land

defense-in-depth   the philosophy behind the design and construction of nuclear plants that calls for overlapping layers of protective features to prevent accidents or to mitigate their effects

detector   a device used to detect the presence of radiation

deuterium   an isotope of hydrogen; its nucleus contains one neutron and one proton

deuteron   the nucleus of a deuterium atom, containing one proton and one neutron

dose   the amount of energy received from radiation by an individual, measured in energy deposited per mass of tissue; usually expressed in rads or grays

dose, absorbed   the amount of energy delivered by ionizing radiation to a material, expressed in rads or grays

dose equivalent   the radiation dose multiplied by a factor based on the relative biological effectiveness of the radiation; usually expressed in rems or sieverts

dose rate   the dose of ionizing radiation delivered per unit of time

dosimeter   a device that measures the total dose of radiation received in a certain period of time

electrical generator   a device that converts mechanical energy into electrical energy

electromagnetic radiation	energy made up of oscillating electric and magnetic fields generated outward from a source; examples of electromagnetic radiation include gamma rays, visible light, microwaves, and radio waves
electron	a small, negatively charged particle that is a basic constituent of matter
element	a chemical substance that cannot be broken into simpler substances without changing its chemical properties
emergency core cooling system	emergency systems in a nuclear reactor designed to cool the core in case the regular reactor coolant system fails
emergency feedwater	backup water supply that, during an accident, can be used to provide water to the steam generators of a pressurized water reactor or, in a boiling water reactor, to the core
excursion	a sudden rise in the power level of a nuclear reactor caused by supercriticality
exposure	the total quantity of radiation received at a given point
external radiation	radiation exposure due to a radiation source outside the body
fast neutron	an energetic neutron produced during a fission reaction
feedwater	in a reactor, the water that is converted to steam and is used to power the electric generator turbines
fertile material	a material (such as thorium-232 or uranium-238) that can be converted into an easily fissionable material by the capture of a neutron
film badge	a device containing radiation-sensitive film that is worn by an individual to measure the absorbed radiation dose received by that person
fissile material	material capable of undergoing induced fission regardless of the energy of the neutron inducing the fission

fission	the splitting of a nucleus into parts (fission products), accompanied by the release of energy, gamma radiation, and neutrons
fissionable materials	materials capable of undergoing fission
fission gases	gaseous fission products; in a reactor, these gases are primarily krypton and xenon
fission products	the nuclei formed during a fission reaction and their radioactive decay products
fossil fuels	substances such as petroleum, coal, and natural gas that are composed of hydrocarbons and may be used for fuels
fuel assembly	a grouping of fuel rods; multiple fuel assemblies are used in the reactor core
fuel cycle	all of the processes involved in the mining, production, use, and disposal of nuclear fuel; can include mining, enrichment, fabrication of fuel elements, reprocessing the spent fuel, and waste disposal
fuel pellet	a small, cylindrical piece of uranium fuel
fuel reprocessing	the processing of spent nuclear reactor fuel to recover unused fissionable material
fuel rod	long, thin tubes that hold the fuel for a reactor; the rods are bundled to form a fuel assembly
fusion reaction	a reaction involving the combination of two light nuclei to form a heavier nucleus, accompanied by the release of energy and neutrons or protons
gamma radiation	high energy electromagnetic radiation, commonly produced during fission; a highly penetrating form of radiation
gas-cooled reactor	a nuclear reactor in which the core is cooled by a gas
gases	a basic form of matter; gases are fluids (like air) that can expand to fill any space

generator	a machine that converts mechanical energy into electrical energy
gigawatt	one billion watts
gigawatt hour	one billion watt hours
graphite	a form of carbon, often used as a moderator in nuclear reactors
gray	a unit of absorbed dose of radiation; one gray equals an absorbed dose of one joule/kilogram
greenhouse effect	warming effect caused by atmospheric carbon dioxide trapping thermal radiation from the earth
half-life	the time it takes for one-half of the atoms of a radioactive substance to disintegrate or transform into other substances
head, reactor vessel	the top portion of the reactor pressure vessel, attached with bolts that can be removed to permit access to the core
heat exchanger	an apparatus that transfers heat from one fluid to another fluid or to the atmosphere; the fluids are usually liquids or gases
heavy water	water containing more deuterium atoms than are usually present in ordinary water; it is used as a moderator in certain reactor designs
heavy water–moderated reactor	a reactor in which the moderator is heavy water
high-level waste	highly radioactive materials, including spent reactor fuel and wastes generated during nuclear fuel reprocessing operations; waste requires special handling and isolation from the environment
ion exchange	a chemical reaction in which ions are exchanged between one substance and another
isotope	two or more forms of a particular element having identical atomic numbers but different numbers of neutrons; isotopes have virtually identical chemical properties but usually dramatically different nuclear properties

kilovolt	one thousand volts
kinetic energy	the energy of a body due to its mass and motion
light water	ordinary water
light water reactor	a nuclear reactor that uses ordinary water as a coolant; boiling water reactors and pressurized water reactors are examples of light water reactors
loop	the path followed by the cooling water in a pressurized water reactor, typically through the pressure vessel to the steam generator, then to the coolant pump, and back to the pressure vessel
loss-of-coolant accident (LOCA)	a reactor accident in which coolant is lost from the primary system
low-level waste	waste with low levels of radioactivity; this waste can be generated in industry, medical facilities, research labs, and nuclear facilities and can often be disposed of in special landfills
mass number	the number of protons and neutrons in the nucleus of an atom
megawatt	one million watts
megawatt hour	one million watt hours
metric ton	one metric ton equals 1,000 kilograms (in the international system of measurement) or approximately 2,200 pounds (in the English system)
microcurie	one millionth of a curie
millirem	one thousandth of a rem
mill tailings	radioactive residue left after the milling process to produce yellowcake
moderator	material used to slow down neutrons to increase the chance of fission
monitored storage	the storage of radioactive material with monitoring of its location and activity
monitoring of radiation	checking of air, water, soil, workers, and the like for radiation levels
nanocurie	one billionth of a curie

natural circulation	circulation without the assistance of pumps
neutron	neutral (uncharged) particle in the nucleus of an atom
neutron, thermal	neutron that is at equilibrium with its environment
neutron capture	the absorption of a neutron by a nucleus during a collision accompanied by the subsequent release of radiation
neutron chain reaction	neutron release from one atom causes neutron release in other atoms (and so on), resulting in a chain reaction
neutron flux	the measure of the number of neutrons passing through a given area per second
neutron generation	release of neutrons by a fissile material resulting from fission and the release of additional neutrons
neutron source	a material (such as plutonium or beryllium) that emits neutrons
noble gas	an inert or chemically unreactive gas, such as argon, helium, krypton, radon, or xenon
non-stochastic effect	refers to health effects in which the severity of the health effect depends on the dose; a dose threshold is assumed
nozzle	connecting tube or vent between various parts of a reactor and the piping system
nuclear energy	energy released through nuclear fission or nuclear fusion
nuclear force	the strong force that binds the nucleus together; it works over a short distance
nuclear power plant	facility that generates electrical energy through nuclear fission
nuclear reaction	reaction involving a change in the nucleus (for example, radioactive decay, fission, or fusion)
nuclear waste	radioactive waste from the nuclear fuel cycle or from the use of radioactive materials (such as those used in medicine)

nucleon	a proton or neutron
nucleus	positively charged part of the atom that contains the protons and (except for hydrogen) the neutrons
nuclide	isotope
pellet, fuel	uranium fuel in the form of small pellets of uranium dioxide
periodic table	chart that lists the elements by atomic number and groups them by common characteristics
personnel monitoring	use of meters to determine radiation exposure of people working with reactors or any other equipment that produces radiation
person-rem	the exposure of one individual to one rem of dose—or the exposure of many individuals to a dose so that when the total dose to the population is added up, it equals one rem
photon	electromagnetic energy unit
pile	slang term used for early reactors
plenum	a large open region in a piping system, usually found at the end of a series of constrictions
plutonium (Pu)	heavy element with atomic number 94; it is fissionable, radioactive, and generally man-made
polisher	removes impurities from the water used in and around a reactor
power reactor	reactor that generates electricity (in contrast to a research reactor)
pressure vessel	strong container that encloses the core, moderator, control rods, and the like
pressurized water reactor	power reactor in which water at a high temperature and under high pressure is used to carry the heat away from the core
pressurizer	tank that provides flexibility in the volume of coolant in a pressurized reactor
primary system	part of the cooling system of the reactor, situated within the containment building

proton	positively charged nuclear particle within the nucleus of the atom
quad	a unit of energy equal to one quadrillion British thermal units (Btu) of energy
quantum theory	concept in physics that energy is radiated in distinct units called "quanta"
rad	a unit of absorbed dose of radiation (acronym for "radiation absorbed dose")
radiation, ionizing	particles that can produce ions (such as alpha particles, beta particles, gamma rays, neutrons, and X rays)
radiation, nuclear	decay from an unstable radioactive nucleus in the form of alpha, beta, and gamma radiation and neutrons
radiation shielding	reducing radiation by the use of a material between the source of radiation and people, equipment, and so on
radioactive contamination	radioactive material in an area that makes that area unsafe
radioactive decay	radiation emitted from unstable atoms, resulting over time in a decrease in radiation (see half-life)
radioactivity	emission of radiation from an unstable nucleus
radioisotope	unstable isotope that emits radiation as it decays
radiological survey	evaluation of radiation hazards
radionuclide	a radioisotope
radium (Ra)	element with atomic number 88 (often found with uranium)
radon (Rn)	element with atomic number 86; a radioactive gas
reactivity	describes the tendency of a reactor to increase or decrease its power
reactor, nuclear	part of a nuclear power plant in which the fission is contained and sustained
reactor building	contains the reactor vessel and associated systems

reactor coolant system	that part of a nuclear plant that removes heat from the reactor core and takes it to the steam turbine
reactor vessel	steel structure around the core of a reactor
reflector	material (such as water, beryllium, or graphite) that surrounds the reactor core and reflects neutrons back to the core to improve efficiency
rem	acronym for "roentgen equivalent man," a standard unit that measures effects of radiation on people
renewable resources	energy sources that regenerate themselves, including solar, wind, and moving water
roentgen (R)	unit of measuring radiation exposure
safeguards	regulations by which materials are controlled and accounted for
safe storage mode	a condition in which a system, such as a nuclear system, is considered to be in a benign non-reactive condition
scabbling	a process by which the surface of a material is removed, leaving the bulk material intact
scattered radiation	a form of secondary radiation in which the direction of the energy has been changed
scintillation detector	detects radiation by a process in which the radiation interacts with a material to produce light
scram	also called a "reactor trip," it is a rapid shutdown of a reactor, usually by the insertion of the control rods; acronym for "safety control rod axe man," from the first nuclear pile in the United States, in which a man with an axe stood ready to cut the ropes holding neutron absorber materials out of the reactor
shielding	protective materials to absorb radiation
shutdown	decrease in the rate of reaction and heat production in a reactor, usually using the control rods
sievert (Sv)	international unit for radiation absorbed; one Sv = one hundred rem

somatic effects of radiation	radiation effects on an individual (as contrasted with genetic effects)
spent fuel	nuclear fuel that is depleted and no longer capable of an efficient sustained chain reaction
spent fuel pool	place to store spent fuel; water supplies some shielding and cooling
stable isotope	an isotope that does not decay radioactively
startup	increase in the fission rate and heat production in a reactor, usually by the removal of control rods
steam generator	a device by which heat from the primary loop is transferred to the secondary loop, which drives the turbines
stochastic effects	effects that may or may not occur after exposure (that is, independent of dose), including cancer and mutation
subcritical	rate of neutron production is decreasing
supercritical	rate of neutron production is increasing
terrestrial radiation	background radiation from normally occurring elements, such as uranium and radon
thermal breeder reactor	breeder reactor utilizing slow (thermal) neutrons
thermal efficiency	ratio of work performed in comparison to the thermal energy input
thermalization	loss of energy by fast neutrons during collisions
thermal power	core heat transfer rate to coolant
thermoluminescent dosimeter	measures radiation by determining the amount of light emitted from a crystal when it is exposed to radiation
thermonuclear	term used to describe a process by which high temperatures are used to fuse light nuclei, resulting in the release of energy
transient	describes a change in the output of a reactor causing a change or resulting from a change in the pressure and/or temperature of the coolant

trip, reactor	a scram
tritium	radioactive isotope of hydrogen with one proton and two neutrons
turbine	rotary engine that turns an electrical generator
ultraviolet	electromagnetic radiation with wavelengths between visible violet and X rays
unstable isotope	radioactive isotope
uranium	radioactive element with atomic number 92; isotope uranium-235 is a fissile fuel used in many reactors
vapor	gaseous form of a substance
vessel head	removable top part of the vessel; it is bolted to the vessel and removed for refueling
waste, radioactive	no longer useful radioactive materials that are ready for storage and disposal
watt	unit of electrical power equal to one joule per second
watt hour	unit of electrical power equal to one watt delivered continuously for one hour
whole-body exposure	radiation distributed relatively uniformly to the body, rather than concentrated in one area (as in radiation treatment)
X rays	photon or electromagnetic radiation with wavelengths less than visible light
yellowcake	a product of the uranium milling process that ranges in color from yellow to dark green, depending on the temperature at which it is dried; also called U_3O_8

Appendix 3: Three Mile Island Unit 2 Fast Facts

Number of reactors at the site: 2

Number of control rooms at the site: 2 (1 per reactor)

Total cost of building Unit 2: $700 million

Amount of concrete in Unit 2: 190,000 cubic yards

Amount of steel (reinforcing and structural): 24,000 tons

Amount of electrical wiring: 740 miles

Amount of electricity produced by Units 1 and 2 when both were in operation: 906 megawatts, sufficient to supply nearly a quarter-million homes

Amount of uranium consumed by Unit 2: 6.07 pounds per day of operation, equivalent to 36,420 barrels of oil or 7,760 tons of coal

Amount of fuel present at the time of the accident: 100 tons, primarily uranium oxide and zirconium

Number of fuel assemblies: 177

Number of fuel rods: 36,816

Amount of water in the cooling loop: 60,000 gallons

Number of fuel casks needed to remove fuel: 22

Fuel in reactor systems after cleanup: 1,100 kilograms (1% of original fuel)

Amount of tritium-contaminated water evaporated: 2.233 million gallons

Cost of the cleanup: $973 million

Estimated average radiation dose to residents: 1 millirem

Estimated average radiation dose to residents within 50-mile radius: 1.5 millirem

Estimated average radiation dose to residents within 5-mile radius: 9 millirem

Average estimated radiation dose from background radiation per year: 100 millirem

Number of radiation dosimeters before the accident: 20

Number of radiation monitoring stations on- and off-site immediately after the accident: 57

Appendix 4: Common Misconceptions About TMI

The accident at TMI contaminated the area around the nuclear reactor.

On the contrary, samples of water, soil, and foodstuffs showed no contamination by radioactive material released from the damaged reactor. Unlike the accident at the Chernobyl reactor in Ukraine, the radioactive material released during the accident was mostly a radioactive form of the noble gases xenon and krypton. These gases do not react chemically and are dispersed quickly in the atmosphere.

Because of the accident, residents of the TMI area have a significantly higher chance of developing health problems than people who reside elsewhere.

Long-term studies of samples of people who resided in the area around TMI during the accident have shown no long-term health effects. Although some researchers have suggested that there may be an elevated incidence of breast cancer, the data are still not conclusive and require further study and follow-up of these residents. Probably the most significant effect was the fear-induced stress resulting from concerns about the accident and conflicting information about the dangers.

No new nuclear power plants were built in the United States as a result of the TMI accident.

While the accident heightened concern about and public opposition to nuclear power, the principal reason that no new nuclear power plants were ordered—and many existing orders cancelled—was an over-supply of generating capacity. Utilities had overbuilt in anticipation of a dramatic growth in consumption that never developed. The growth rate in electrical consumption of the 1960s and early 1970s was 8% per year; it dropped in the 1970s to 1–2% per year, resulting in too much generating capacity. Plans for both new nuclear and new coal-fired power plants were cancelled as a result of this drop. As of the writing of this book, certain areas of the United States are experiencing power shortages, and utilities are building new power plants. These new plants are either combined-cycle gas or gas turbine plants. For a variety of reasons, long-term considerations are being given to new nuclear power plants.

The accident at TMI had a significant impact on the environment.

Unlike the accident at the Chernobyl reactor, which contaminated a large area around the reactor, the accident at TMI did not release the types of material that can combine and remain in the soil or water for long periods of time. Only radioactive forms of xenon and krypton were released, and these elements are noble gases that do not react chemically.

A nuclear power plant could blow up like an atomic bomb.

The nuclear fuel in a commercial power reactor is simply not con-centrated enough to explode like a bomb. The amount of uranium-235 (a uranium isotope) is typically no more then 5% of the fuel. In a nuclear weapon, the amount of such material is greater than 90%.

There were many mutations in plants and animals in the area around TMI as a result of the accident.

Long-term environmental and health studies have found no genetic effects from the accident.

Many people were killed as a result of the TMI accident.

No one was killed as a result of the accident. The maximum radiation exposure to anyone who remained outdoors in the area of the plant during the entire accident was less then the average background radiation levels in the Harrisburg area. The average dose to people living within fifty miles of the plant was about the same as the amount some-one receives from natural background radiation by flying at 30,000 feet from New York City to Kansas City.

Appendix 5: Acronyms

ACRS

The Advisory Committee on Reactor Safeguards was mandated by the Atomic Energy Act of 1954. Its charge is "to review and report on safety studies and reactor facility license and license renewal applications; to advise the Commission on the hazards of proposed and existing reactor facilities and the adequacy of proposed reactor safety standards; and to initiate reviews of specific generic matters or nuclear facility safety–related items." The ACRS Web site can be found at http://www.nrc.gov/what-we-do/regulatory/advisory/acrs.html.

AEC

The Atomic Energy Commission was established in 1946 for the civilian development of atomic energy. It was abolished in January 1975 and replaced in part by the Energy Research and Development Administration (ERDA).

BRP

The Pennsylvania Bureau of Radiation Protection is the state organization that regulates and monitors all types of man-made and naturally occurring sources of radiation in the Commonwealth. The BRP Web site can be found at http://www.dep.state.pa.us/dep/deputate/airwaste/rp/rp.htm.

CFR

The Code of Federal Regulations is the collection of rules from the executive departments and agencies of the federal government. The CFR Web site can be found at http://www.gpoaccess.gov/cfr/index.html.

DOE

The U.S. Department of Energy is the executive agency formed from the Federal Energy Administration and the Energy Research and Development Administration in 1977. The DOE Web site can be found at www.energy.gov.

EIA

The Energy Information Administration is the statistical branch of the Department of Energy. The EIA Web site can be found at http://www.eia.doe.gov.

ERDA

The Energy Research and Development Administration was created from the AEC in 1974. It was closed in 1977 after the formation of the DOE.

INPO

The Institute of Nuclear Power Operations was created by the U.S. nuclear industry in 1980 to improve the safety and reliability of nuclear plants and the training of reactor operators.

NRC

The Nuclear Regulatory Commission was created in 1974 from the AEC to regulate nuclear fuels and facilities and to safeguard public health and safety. The NRC Web site can be found at www.nrc.gov.

PEMA

The Pennsylvania Emergency Management Agency is the agency charged with coordinating response to man-made and natural disasters. The PEMA Web site can be found at http://www.pema.state.pa.us.

Bibliography

American Chemical Society. 1986. *The Three Mile Island Accident: Diagnosis and Prognosis*. Edited by L. M. Toth et al. Washington, DC: American Chemical Society.

Arnold, Lorna. 1992. *Windscale 1957: Anatomy of a Nuclear Accident*. Basingstoke, Hampshire: Macmillan.

Birnie, Jill Anne. 1982. "Three Mile Island: A Study of Credibility in a Crisis." Ph.D. diss., University of Pittsburgh.

Bromet, Evelyn J., David K. Parkinson, Herbert C. Schulberg, Leslie O. Dunn, and Paul C. Gondek. 1982. "Mental Health of Residents near the Three Mile Island Reactor: A Comparative Study of Selected Groups." *Journal of Preventive Psychiatry* 3 (1): 225–76.

Broughton, J. M., K. Pui, D. A. Petti, and E. L. Tolman. 1989. "A Scenario of the Three Mile Island Unit 2 Accident." *Nuclear Technology* 87 (1): 34–53.

Cable, Sherry, Edward J. Walsh, and Rex H. Warland. 1988. "Differential Paths to Political Activism: Comparisons of Four Mobilization Processes After the Three Mile Island Accident." *Social Forces* 66 (4): 951–69.

Cantelon, Philip L., and Robert C. Williams. 1982. *Crisis Contained: The Department of Energy at Three Mile Island*. Carbondale: Southern Illinois University Press.

Davidson, Laura M., Raymond Fleming, and Andrew Baum. 1987. "Chronic Stress, Catecholamines, and Sleep Disturbance at Three Mile Island." *Journal of Human Stress* 13 (2): 75–82.

Department of Energy. 1987. *Atoms to Electricity*. DOE/NE-0085. Washington, DC: U.S. Department of Energy.

Dew, M. A., and E. J. Bromet. 1993. "Predictors of Temporal Patterns of Psychiatric Distress During 10 Years Following the Nuclear Accident at Three Mile Island." *Social Psychiatry and Psychiatric Epidemiology* 28 (3): 49–50.

Dew, Mary Amanda, Evelyn J. Bromet, and Herbert C. Schulberg. 1987. "A Comparative Analysis of Two Community Stressors' Long-term Mental Health Effects." *American Journal of Community Psychology* 15 (2): 167–83.

Dew, Mary Amanda, Evelyn J. Bromet, Herbert C. Schulberg, Leslie O. Dunn, and David K. Parkinson. 1987. "Mental Health Effects of the Three Mile Island Nuclear Reactor Restart." *American Journal of Psychiatry* 144 (8): 1074–77.

Dimopoulos, Kostas, and Vasilis Kouladis. 2002. "Science and Technology Education for Citizenship: The Potential Role of the Press." *Science Education* 87 (3): 241–56.

Dresser, Peter D., ed. 1993. *Nuclear Power Plants Worldwide*. Detroit: Gale Research.

Eisenbud, Merril. 1989. "Exposure of the General Public near Three Mile Island." *Nuclear Technology* 87 (2): 514–19.

Electric Power Research Institute. 1980. *Analysis of the Three Mile Island– Unit 2 Accident*. NSAC-8C-1 (NSAC-1 Revised). Palo Alto, CA: Electric Power Research Institute, Nuclear Safety Analysis Center.

Energy Information Administration. 2003a. *Annual Energy Outlook 2003 with Projections to 2025*. DOE/EIA-0383. Washington, DC: U.S. Department of Energy, Energy Information Administration. Available online at http://www.eia.doe.gov/oiaf/aeo/.

———. 2003b. *International Energy Outlook 2003*. DOE/EIA-0484. Washington, DC: U.S. Department of Energy, Energy Information Administration. Available online at http://www.eia.doe.gov/ oiaf/ieo/.

Environmental Protection Agency. 1980. *Long-Term Environmental Radiation Surveillance Plan for Three Mile Island*. Washington, DC: U.S. Environmental Protection Agency.

Farrell, Thomas B., and G. Thomas Goodnight. 1981. "Accidental Rhetoric: The Root Metaphors of Three Mile Island." *Communication Monographs* 48 (4): 271–300.

Field, R. William. 1993. "^{137}Cs Levels in Deer Following the Three Mile Island Accident." *Health Physics* 64 (6): 671–74.

Field, R. William, et al. 1981. "Iodine-131 in Thyroids of the Meadow Vole *(Microtus pennsylvanicus)* in the Vicinity of the Three Mile Island Nuclear Generating Plant." *Health Physics* 41 (8): 297–301.

Friedman, Sharon M. 1981. "Blueprint for Breakdown: Three Mile Island and the Media Before the Accident." *Journal of Communication* 31 (2): 116–28.

Friedman, Sharon M., Carole M. Gorney, and Brenda P. Egolf. 1987. "Reporting on Radiation: A Content Analysis of Chernobyl Coverage." *Journal of Communication* 37 (3): 58–67.

Gallup, George H. 1978. *The Gallup Poll: Public Opinion, 1972–1977*. Wilmington, DE: Scholarly Resources.

———. 1980. *The Gallup Poll: Public Opinion, 1979*. Wilmington, DE: Scholarly Resources.

Gatchel, Robert J., Marc A. Schaeffer, and Andrew Baum. 1985. "A Psycho-physiological Field Study of Stress at Three Mile Island." *Psychophysiology* 22 (2): 175–81.

Gears, G. E., G. LaRoche, J. Cable, B. Jaroslaw, and D. Smith. 1980. *Investigation of Reported Plant and Animal Health Effects in the Three Mile Island Area.* NUREG-0738. Washington, DC: U.S. Nuclear Regulatory Commission.

General Public Utilities Nuclear Corporation. N.d. "TMI and Nuclear Power Facts and Figures." Parsippany, NJ: GPU Nuclear Corporation.

———. 1982. TMI-2 Quick Look TV Cam Inspection #1, Edited. Videocassette V1071. Middletown, PA: General Public Utilities Corporation. Available at The Pennsylvania State University TMI-2 Recovery and Decontamination Collection, University Libraries, University Park, PA.

Goldstein, Raymond, John K. Schorr, and Karen S. Goldsteen. 1989. "Longitudinal Study of Appraisal at Three Mile Island: Implications for Life Event Research." *Social Science and Medicine* 28 (4): 389–97.

Good, Beverly A., Gordon M. Lodde, and Diane M. Surgeoner. 1989. "Three Mile Island and the Environment." *Nuclear Technology* 87 (2): 395–406.

Gur, David, W. F. Good, G. K. Tokuhata, M. K. Goldhaber, J. C. Rosen, G. V. Rao, J. M. Herron, D. M. Miller, and R. S. Hollis. 1983. "Radiation Dose Assignment to Individuals Residing near the Three Mile Island Nuclear Station." *Proceedings of the Pennsylvania Academy of Science* 57: 99–102.

Hatch, M. C., S. Wallenstein, J. Beyea, J. W. Nieves, and M. Susser. 1991. "Cancer Rates After the Three Mile Island Nuclear Accident and Proximity of Residence to the Plant." *American Journal of Public Health* 81 (6): 719–24.

Hofstetter, K. J., and B. S. Asmus. 1989. "The Identification and Control of Microorganisms at Three Mile Island Unit 2." *Nuclear Technology* 87 (4): 837–44.

Hohenemser, Christoph, and Ortwin Renn. 1988. "Chernobyl's Other Legacy." *Environment* 30 (3): 4–11, 40–45.

Holton, W. C., C. A. Negin, and S. L. Owrutsky. 1990. *The Cleanup of Three Mile Island Unit 2—A Technical History: 1979–1990.* EPRI NP-6931. Palo Alto, CA: Electric Power Research Institute.

Houts, Peter S., Paul D. Cleary, and The-Wei Hu. 1988. *The Three Mile Island Crisis: Psychological, Social, and Economic Impacts on the Surrounding Population.* University Park: The Pennsylvania State University.

Hull, Andrew P. 1989. "Environmental Measurements During the Three Mile Island Unit 2 Accident." *Nuclear Technology* 87 (2): 383–94.

Institute of Nuclear Power Operations. 2003. *Performance Indicators for the U.S. Nuclear Industry.* Atlanta: Institute of Nuclear Power Operations. Available online at http://www.nei.org/documents/Wano_Performance _Indicators_2002.pdf.

Keisling, Bill. 1980. *Three Mile Island: Turning Point.* Seattle: Veritas Books.

Kemeny, J. G., et al. 1979. *Report of the President's Commission on the Accident at Three Mile Island—The Need for Change: The Legacy of TMI.* Washington, DC: U.S. Government Printing Office.

Kerecz, John J. 1986. "Background Gamma Radiation Levels in the Three Mile Island Area Since the Accident in 1979." M.E.P.C. thesis, The Pennsylvania State University.

Khan, Rahat Nabi. 1988. "Science, Scientists, and Society: Public Attitudes Towards Science and Technology." *Impact of Science on Society* 151: 257–71.

Lanouette, William. 1989a. *The Atom, Politics, and the Press.* Washington, DC: Woodrow Wilson International Center for Scholars.

———. 1989b. *Three Mile Island +10: Will Press Coverage Be Better Next Time?* Washington, DC: Woodrow Wilson International Center for Scholars.

MacGregor, Donald. 1991. "Worry over Technological Activities and Life Concerns." *Risk Analysis* 11 (2): 315–24.

Martin, Daniel. 1980. *Three Mile Island: Prologue or Epilogue?* Cambridge, MA: Ballinger.

Media Institute. 1979. *Television Evening News Covers Nuclear Energy.* Washington, DC: The Media Institute.

———. 1980. *The Public's Right to Know: Communicators' Response to the Kemeny Commission Report.* Washington, DC: The Media Institute.

Meighan, Don G. 1976. "How Safe Is Safe Enough?" *New York Times Magazine,* June 20, 173–75.

Morse, Roger A., Darrell R. Van Campen, Walter H. Gutenmann, Donald J. Lisk, and Clarence Collison. 1980. "Analysis of Radioactivity in Honeys Produced near Three-Mile Island Nuclear Power Plant." *Nutrition Reports International* 22 (3): 319–21.

National Cancer Institute. 1979. *Known Health Effects of Low-Level Radiation Exposure: Health Implications of the TMI (Three Mile Island) Accident.* DHEW/PUB/NIH80-2087. Bethesda, MD: National Cancer Institute.

National Council on Radiation Protection and Measurements. 1987. *Exposure of the Population in the United States and Canada from Natural Background Radiation.* Bethesda, MD: National Council on Radiation Protection and Measurements.

National Research Council. 1990. *Health Effects of Exposure to Low Levels of Ionizing Radiation.* Washington, DC: National Academy Press.

————. 2002. *Technically Speaking: Why All Americans Need to Know More About Technology.* Edited by Greg Pearson and A. Thomas Young. Washington, DC: National Research Council.

Nimmo, Dan, and James E. Combs. 1982. "Fantasies and Melodramas in Television Network News: The Case of Three Mile Island." *The Western Journal of Speech Communications* 46 (Winter): 45–55.

Nuclear Energy Agency. 2003. "Chernobyl: Assessment of Radiological and Health Impacts: 2002 Update of Chernobyl: Ten Years On." http://www.nea.fr/html/rp/chernobyl/chernobyl.html.

Nuclear Energy Institute. 2003. "Comparative Measures of Power Plant Efficiency." http://www.nei.org/doc.asp?docid=1064.

Nuclear Regulatory Commission. 1980a. *Final Environmental Assessment for Decontamination of the Three Mile Island Unit 2 Reactor Building Atmosphere.* 2 vols. NUREG-0662. Washington, DC: U.S. Nuclear Regulatory Commission.

————. 1980b. *NRC Action Plan Developed as a Result of the TMI-2 Accident.* 2 vols. NUREG-0660. Washington, DC: U.S. Nuclear Regulatory Commission.

————. 1983. *Programmatic Environmental Impact Statement (Final) Related to Decontamination and Disposal of Radioactive Wastes Resulting from March 28, 1979 Accident Three Mile Island Station, Unit 2.* NUREG-0683. Washington, DC: U.S. Nuclear Regulatory Commission.

————. Ad Hoc Interagency Population Dose Assessment Group. 1979. *Population Dose and Health Impact of the Accident at the Three Mile Island Nuclear Station.* NUREG-0558. Washington, DC: U.S. Government Printing Office.

Nuclear Safety Analysis Center. 1981. *TMI-2 Accident Core Heat-Up Analysis: A Supplement.* NSAC-25. Palo Alto, CA: Nuclear Safety Analysis Center.

Patterson, Philip Don. 1987. "Nuclear Networks: How Television News Covers Technological Crises." Ph.D. diss., University of Oklahoma.

Pearson, Greg. 2002. "What Americans Know (or Think They Know) About Technology." *Issues in Science and Technology* 18 (4): 80–82.

Rees, Joseph V. 1994. *Hostages of Each Other: The Transformation of Nuclear Safety Since Three Mile Island.* Chicago: University of Chicago Press.

Rogovin, M., et al. 1980. *Three Mile Island: A Report to the Commissioners and to the Public.* 2 vols. NUREG/CR-1250. Washington, DC: U.S. Nuclear Regulatory Commission.

Rolph, Elizabeth S. 1979. *Nuclear Power and the Public Safety.* Lexington, MA: Lexington Books.

Rubin, David M. 1987. "How the News Media Reported on Three Mile Island and Chernobyl." *Journal of Communication* 37 (3): 42–57.

Schaeffer, Marc A., and Andrew Baum. 1984. "Adrenal Cortical Response to Stress at Three Mile Island." *Psychosomatic Medicine* 46 (3): 227–35.

Sorensen, John, et al. 1987. *Impacts of Hazardous Technology: The Psychosocial Effects of Restarting TMI-1.* Albany: State University of New York Press.

Stephens, Mark. 1980. *Three Mile Island.* New York: Random House.

Stephens, Mitchell, and Nadyne G. Edison. 1982. "News Media Coverage of Issues During the Accident at Three Mile Island." *Journalism Quarterly* 59 (Summer): 199–204+.

Talbott, E. O., A. O. Youk, K. P. McHugh, J. D. Shire, A. Zhang, B. P. Murphy, and R. A. Engberg. 2000. "Mortality Among the Residents of the Three Mile Island Accident Area: 1979–1992." *Environmental Health Perspectives* 108 (6): 545–52.

Talbott, E. O., A. O. Youk, K. P. McHugh-Pemu, and J. V. Zborowski. 2003. "Long-term Follow-up of the Residents of the Three Mile Island Accident Area: 1979–1998." *Environmental Health Perspectives* 111 (3): 341–48.

Thomas, P. J., and R. Zwissler. 2003. "New Predictions for Chernobyl Childhood Thyroid Cancers." *Nuclear Energy* 42 (4): 203–11.

Tromp, T. K., R.-L. Shia, M. Allen, J. M. Eiler, and Y. L. Yung. 2003. "Potential Environmental Impact of a Hydrogen Economy on the Stratosphere." *Science* 300 (5626): 1740–42.

United Nations. 1997. *Kyoto Protocol to the United Nations Framework Convention on Climate Change.* New York: United Nations. Also available online at http://unfccc.int/resource/convkp.html.

Walker, J. Samuel. 1992. *Containing the Atom: Nuclear Regulation in a Changing Environment, 1963–1971.* Berkeley and Los Angeles: University of California Press.

Walsh, Edward J. 1988. *Democracy in the Shadows: Citizen Mobilization in the Wake of the Accident at Three Mile Island.* New York: Greenwood.

Wasserman, Harvey. 1996. "In the Dead Zone: Aftermath of the Apocalypse." *The Nation* 262 (17): 16–20.

Wing, S., D. Richardson, D. Armstrong, and D. Crawford-Brown. 1997. "A Reevaluation of Cancer Incidence near the Three Mile Island Nuclear Plant: The Collision of Evidence and Assumptions." *Environmental Health Perspectives* 105 (1): 52–57.

Ziman, John. 1991. "Public Understanding of Science." *Science, Technology, and Human Values* 16 (1): 99–105.

Index

ABC (American Broadcasting Company), 53
acid rain, 96
Action Plan (NRC), 88–89
Ad Hoc Population Dose Assessment Group, 66, 68
adrenal cortical activity, 72
advanced boiling water reactors, 10
advanced light water reactors, 8, 10
advanced pressurized water reactor, 109
Advisory Committee on Reactor Safeguards, 78, 79
air pollution, 95
alarms, 22, 23, 26
alpha particles, 60
alternative energy resources, 100–106
American Nuclear Insurers, 35
Annual Energy Outlook, 93
anticontamination clothing, 38
antinuclear protest groups. *See* intervenor groups
Apollo 1, 115
Apollo 13, 115
AP600 reactor, 10, 109
Arab oil embargo of 1973, 76
Arab oil embargo of 1979, 92
Associated Press, ix, x
atom
    and radioactive decay, 60–61
    splitting, 2–5
Atomic Energy Act of 1946, 76
Atomic Energy Act of 1954, 77

Atomic Energy Commission, 13, 76–77, 78, 79
atomic number, 3
Atoms for Peace, 13
auxiliary building, 37, 38, 48
    ventiliation, 25, 26

Babcock and Wilcox, 15, 31, 87
background radiation, 62, 69
backup systems, 23
becquerels, 62
Beers, Paul, 52
Belgium, 110
biomass, 1, 102, 105–6
Black Friday, 30
Bohr, Niels, 50
boiling water reactors, 8–9, 17
boric acid, 12
boring equipment, 41
boron, 12
breeder reactors, 8, 10, 102, 108
British thermal unit, 91
Brown's Ferry Nuclear Power Plant, 17
"bug kill," 45
Bulgaria, 110
Burns and Roe, 15

cancer, 18, 27, 61, 65, 66, 68–70, 75
CANDU reactor, 10
capacity factor, 90
carbon, 108
carbon dioxide, 10, 95–96, 97, 102, 108, 112

carbon monoxide, 95
Carter, Jimmy, 30, 80, 83
Carter, Rosalynn, 30
casks, 7
CBS (Columbia Broadcasting
    System), 53
cesium, 31, 48, 74, 75
    cesium-137, 38, 75
Chernobyl, 17, 18–19, 31, 48, 55,
    117
China, 95, 110
China syndrome, 47, 48
China Syndrome, The, xi, 19
chromosomes, 65
    mutations in, 65
cladding, 12, 47
Clamshell Alliance, 56
cleanup, 33–48
    cost, 85
CNN, 49
coal, 92, 94, 95, 102, 112
    reserves, 102
cobalt-60, 74
Code of Federal Regulations, 75, 79
cogeneration, 11, 98
cold shutdown, 33
coliforms, 44
collecting tank, 23, 26, 38
Columbia University, 68
combustion, 95
condenser, 9, 11
containment building, xi, 12, 18, 39,
    48
    decontamination, 36, 42
    recontamination, 42, 43, 44
    venting, 28, 39
containment purging, 39
control rods, 6, 7, 12, 22
control room, ix, 11, 12, 15, 22, 52
    design, 15, 32, 87, 89, 90, 114
coolants, 7, 24, 25
    air, 7
    carbon dioxide, 10

heavy water, 7
    helium, 7, 10
    liquid sodium, 7
    liquid sodium–potassium alloy, 7
    water, 7
cooling tower, 1, 10–11
core, xi, 5, 6, 7, 9, 12, 17, 18, 25, 42,
    47
    bores, 41
    cooldown, 47
    debris bed, 47
    defueling, 44–46
    rubble bed, 45
    sampling, 41
critical state, 7
curie, 62
Curie, Marie, 13

dairies, 73
Dartmouth College, 81
Davis-Besse, 31
Decatur, Alabama, 17
decay heat, 12–13
decommissioning, 13, 17, 37, 46
decontamination, 36, 38–40, 42–43
    EPICOR, 38
    floor scabbling, 38
    ion exchanger, 38
    power washing, 38
    robotic surveys, 42
    steam/vacuum surface cleaning, 38
    strippable coating, 38
    submerged demineralizer system,
        38
    water lancing, 38
defueling, 39, 43–46
    dry defueling, 44
    refueling canal, 39, 43–44
    rubble bed, 44
    shipping casks, 46
    tools, 43, 44, 45
    work platform, 42, 44
Denton, Harold, 30, 34, 51

deuterium, 111
Dresden Unit 1, 14
dress out, 38
drills, 89
dry defueling, 44
Duquesne Light Company, 13, 78

Einstein, Albert, 13
Eisenhower, Dwight D., 13
electricity demand, 107
Electric Power Research Institute, 39
electrons, 2
Elizabethtown, Pennsylvania, 34
emergency core cooling system, 23,
    27, 28, 47
energy consumption, 92–95, 112
Energy Information Administration,
    92
energy projections, 93
Energy Reorganization Act of 1974,
    80
Energy Research and Development
    Administration, 13, 80
energy reserves, 99, 112
energy sources
    biomass, 1, 102, 105
    coal, 92, 94, 102, 112
    consumption, 141
    geothermal, 1, 102, 106
    hydropower, 1, 92, 95
    natural gas, 92, 94, 95, 97, 101,
        107
    nuclear, 102, 107–10, 112
    ocean, 102, 106
    oil, 94, 100
    solar, 1, 95, 102, 103–4
    tidal, 106
    wind, 1, 92, 95, 102, 104
environment, 95
    carbon dioxide and, 73–75, 95, 96,
        97, 102, 108, 112
    carbon monoxide and, 95

effects of TMI 2 on, xiii, 31, 48, 73
    global warming and, 96
    greenhouse effect and, 96
    greenhouse gases and, 96, 97
    ozone and, 95, 99
    particulates in, 95
    sulfur dioxide and, 108
EPICOR, 46
EPICOR I, 38
EPICOR II, 38
equipment design, 32
erythema, 65
ethanol, 106
evacuation advisory, 29
evaporation, radioactive water, 46

fast breeder reactors, 8
feedwater system, 22
Fermi, Enrico, 50
fertile material, 6
fishing, 73
fissile material, 5
    plutonium-239, 5, 10, 108
    uranium-235, 5–6
fission, 3, 4, 12
    fertile material, 6
    fissile material, 5, 10
    fission fragments, 12
    plutonium, 17, 108
    plutonium-239, 5
    uranium-235, 5–6
floor scabbling, 38
Forked River, New Jersey, 15
France, 85, 110
Friends of the Earth, 56
fuel, 26, 41, 45
    assembly, 6, 7
    cells, 97
    pellets, 6, 8, 12, 110
    removal, 39, 42
    rods, 1, 6, 7, 12, 25, 30, 44, 108
fusion energy, 111

gamma radiation, 60, 65
gas-cooled fast breeder reactor, 10
gas-cooled thermal reactors, 8, 10
General Electric, 76
General Public Utilities, ix, 35, 39, 82, 85, 86, 87
Generation IV reactors, 110
generator, 5, 11
geothermal energy, 1, 102, 106
Gilbert Associates, 15
global warming, 96
Goldsboro, Pennsylvania, 26, 27
"go solid," 24
green energy resources, 103
greenhouse effect, 96
greenhouse gases, 96, 97, 108
Greenpeace, 18

hafnium, 12
half-life, 61
Harrisburg, viii
Harrisburg Patriot, viii, xi, xii, 52
health effects
    cancer, 18, 27, 61, 65, 66, 68–70
    mortality studies, 69, 70
    sleep disturbances, 71
    stress, 72, 75
heart disease, 69, 70
heavy water, 10
heavy water moderated reactors, 8
helium, 5, 7, 10, 18
Henry, Thomas R., 50
Herbein, John, x
Hershey Foods Corporation, viii
high temperature gas-cooled reactors, 8, 10
high temperature nuclear reactors, 97
Hungary, 110
hydrogen, 25, 26, 27, 29, 30, 33, 60, 97
hydrogen bubble, xii, 30
hydrogen burn, 37, 43
hydrogen economy, 96–99

hydrogen isotopes, 3
hydrogen peroxide, 45
hydropower, 1, 92, 95

Idaho National Engineering Laboratory, 44, 46
India, 95, 110
Indian Point Nuclear Generating Station, 56
inertial confinement, 111
inspections, 89
Institute of Nuclear Power Operations, 88
International Atomic Energy Agency, 18
International Technology Education Association, 58
intervenor groups
    Clamshell Alliance, 56
    Mobilization for Survival, 56
    National Critical Mass, 56
    Newberry Township Steering Committee, 56
    PANE, 35, 56
    Susquehanna Valley Alliance, 56
    TMI Alert, 56
iodine, 27, 28, 31, 74
    iodine-131, 17, 38, 65, 74
    iodine-133, 65
ion exchangers, 38
isotopes, 3, 60, 62, 108
isotopes, radioactive, 73, 108

Japan, 85, 110
Jersey Central Power and Light Company, ix
Joint Committee on Atomic Energy, 84

Kemeny, John G., 81
Kemeny Commission. *See* President's Commission on the Accident at Three Mile Island

krypton, 33–34
  krypton-85, 33–34, 39, 65
Kyoto Protocol, 96

lawsuits, 86–87
lead, 61
leukemia, 69
Liberty Fire Hall (Middletown, Pa.),
  34
light water breeder reactors, 10
light water reactors, 8
linear model, 64
liquid metal–cooled fast breeder
  reactors, 10
Lithuania, 110
livestock, 73
loss-of-coolant accident, 24, 25, 31,
  79
Love Canal, 117

magnetic confinement, 111
main steam line break accident, 24
Manhattan Project, 13, 19
meadow vole, 74
media coverage, x, 49, 51, 82, 84
  newspaper, 54
  radio, 53
  television, 28, 50, 53
Media Institute, 50, 52, 53, 54
methane, 97
  cracking, 97
Metropolitan Edison, ix, x, 27, 34,
  35, 39, 51, 54, 116
microorganisms, 44
Middletown, Pennsylvania, xii, 34
milk, 73
mining, 6, 10
mini-submarine, 45
Mitchell, William, 79
Mobilization for Survival, 56
moderators, 5, 7
  beryllium, 5, 8
  graphite, 5, 8

heavy water, 5, 8
light water, 5, 8
liquid sodium, 8
molten salt breeder, 10
monitoring, 75
mortality rate, 69
mortality studies, 68, 70

National Academy for Nuclear
  Training, 88
National Academy of Engineering,
  58
National Critical Mass, 56
natural gas, 92, 94, 95, 97, 101, 107,
  112
  reserves, 101
NBC (National Broadcasting
  Company), 53
neptunium-239, 10
neutrons, 2, 7, 12
Newberry Township Steering
  Committee, 56
New York Times, xi, xii
NIMBY, 19, 109
nitrogen dioxide, 95
nitrogen oxide, 95, 108
nitrogen-16, 61
nonrenewable energy resources, 1,
  100–102
non-stochastic health effects, 65
nuclear energy, 14, 49, 107, 110
  basics, 1–5
nuclear forces, 3
Nuclear Free Hudson, 56
nuclear industry, 76, 81, 84
  plant cancellations, 85, 90
  plant orders, 78, 79, 85, 112
  turnkey plants, 78
Nuclear Information and Resources
  Service, 56
nuclear power, 92, 94, 107–11
nuclear reactors, 107

Nuclear Regulatory Commission, x, 21, 22, 26, 28, 31, 34, 35, 39, 51, 54, 66, 73, 80–84, 88–89, 90, 109
nucleus, 2

ocean power, 102, 106
oil, 94, 100–101
    reserves, 100
once-through cycle, 7
OPEC, 91
operators, 23–27, 31
Organisation for Economic Co-operation and Development Nuclear Energy Agency, 18
Oyster Creek Nuclear Generating Station Unit 2, 15
ozone, 95, 99

PANE. *See* People Against Nuclear Energy
particulates, 95
peat, 106
pebble bed modular reactor, 110
"peep show," 37
Pennsylvania Bureau of Radiation Protection, 26
Pennsylvania Department of Environmental Resources, 73
Pennsylvania Department of Health, 67
Pennsylvania Electric Company, ix
Pennsylvania Emergency Management Agency, 26
Pennsylvania Public Utility Commission, 35
Pennsylvania State University, The, 39
People Against Nuclear Energy, 35
periodic table, 3
person-rem, 64
petroleum, 92, 95
Philadelphia Inquirer, viii, xii, 34

photons, 104
plasma arc automatic cutting system, 45
plumes, radioactive, 11, 69, 73, 118
plutonium, 3, 10, 108
    plutonium-239, 5, 10, 108
polar crane, 7, 40, 43
polishing system, 22
post-defueling monitored storage, 37, 46
Power Demonstration Reactor Program, 77
power washing, 38
President's Commission on the Accident at Three Mile Island, 52, 54–55, 80, 83, 88
    recommendations, 82
pressure vessel. *See* reactor vessel
pressurized water reactors, 8
Price-Anderson Act (1957), 77, 79
primary cooling system, 24, 29, 45
primary loop, 9
protons, 2
Pseudomonas paucimobilis, 44
psychiatric symptoms, 72
Public Citizen, 56

quad, 92
"Quick Look," 40–41

radiation, 19, 31, 60, 66–67
    alpha particles, 60
    beta particles, 60, 65
    dose, 64, 65
    gamma rays, 60, 70
    levels, 26, 43
    monitors, 28, 73
radioactive decay law, 61–62
radioactive gases, 25, 26
radioactive isotopes, 31, 61
    cesium-137, 38, 75
    iodine-131, 17, 38, 65, 74

radioactive isotopes (*cont.*)
  iodine-133, 65
  krypton-85, 33–34, 39, 65
  radon, 6, 63–64, 69
radioactive wastes, 7, 108–9
radioactivity, xii, 1, 27, 42, 48, 60, 62
radon, 6, 63–64, 69
reactor buildings, 1
reactor coolant system, 23
reactor core, 11
reactor protection system, 12
reactor vessel, 6–7, 9, 43–45
redundant systems, 11, 12
refueling, 6, 7
refueling canal, 39, 43–44
rem, 62
Remote Reconnaissance Vehicle, 42
renewable energy resources, 1, 92,
  102–6
respirators, 38
Rickover, Hyman, 77
river sediments, 74
robotic surveys, 42
Roentgen, Wilhelm, 13
Rogovin, Stern, and Huge, 83, 88
  recommendations, 83
Rover. *See* Remote Reconnaissance
  Vehicle
rubble bed, 44–45

safety features, 11–13
  passive, 10
safety regulations, 84
scientific literacy, 57
scrams, 90
Scranton, William, x, 27
secondary loop, 9
self-contained breathing apparatus,
  38
self-sustaining reaction, 5
shipping casks, 46

Shippingport, Pennsylvania, 2, 13,
  78, 107
Shippingport Atomic Power Station,
  13
Sholly decision, 35
sievert, 62
simplified boiling water reactor, 10
site emergency, 26
Slovakia, 110
solar energy, 1, 95, 102, 103–4
South Korea, 95, 110
statistical significance, 67
steam generator, 9, 45
steam/vacuum surface cleaning, 38
stochastic health effects, 65
stress, 71–72, 75, 81
  chronic, 71
  levels, 68
strippable coating, 38
strontium, 31, 48
subcritical reaction, 5
submerged demineralizer system, 38
sugarcane, 106
sulfur dioxide, 95, 108
supercritical reaction, 5
Susquehanna River, 15, 46, 73, 74,
  75
Susquehanna River Valley, 73
Susquehanna Valley Alliance, 56
System 80+ reactor, 10

Taiwan, 95
technological literacy, 57, 82
Tennessee Valley Authority, 17, 31
thermal reactors, 5
Thornburgh, Richard, 28, 30, 35, 60
Thornburgh plan, 35
Three Mile Island Public Health
  Fund, 87
Three Mile Island Unit 1, 15, 85, 115

Three Mile Island Unit 2, 15–16, 21, 46, 66
  auxiliary building, 25, 26, 27, 31, 38
  containment building, 23, 25, 26, 27, 31, 34, 39, 40
  coolant pumps, 24
  core, 25, 26, 28, 30, 37, 40, 41, 42, 44, 46, 47, 48
  generator, 22
  head, 28, 43
  lower head, 30, 40, 41, 45, 47
  pressurizer, 22, 24
  pumps, 24
  reactor vessel, 40, 41, 43
  steam generator, 24
  testing, 16
  turbines, 24
thyroid, 18, 27, 74
tidal power, 106
TMI cohort, 67, 68, 69, 70
TMI Information and Examination Program, 39
TMI Population Registry, 68
TMI Recovery and Decontamination Database, 118
training, 32
tritium, 3, 46, 74, 111
turbine, 5, 9, 11, 22, 24, 33

Ukraine, 18, 110
Union of Concerned Scientists, 56
United Engineers and Constructors, 15
United Nations, 96
United Nations Scientific Committee on the Effects of Atomic Radiation, 18
University of North Carolina, 68
University of Pittsburgh, 69, 70
upper plenum, 43

uranium, 2, 3–7, 8, 10, 48, 50, 61, 62, 63, 107, 108
  enrichment, 6
  uranium-233, 108
  uranium-235, 5, 6, 10
  uranium-238, 6, 10, 108
U.S. Congress, 84
U.S. Court of Appeals for the Third Circuit, 35, 87
U.S. Department of Energy, 34, 35, 39, 73, 74, 98, 110
  Radiological Assistance Plan Office, 26
U.S. Department of Health, Education, and Welfare, 66
U.S. Environmental Protection Agency, 39, 66, 74
U.S. Food and Drug Administration, 74
U.S. Navy, 77
USS *Nautilus,* 77
U.S. Supreme Court, 35

valves, 16, 22, 23, 24, 26, 27
  backup, 26
  open relief, 46
  power-operated relief, 22, 26, 31
  pressure regulating, 16, 22, 23
  relief, 27
Vanderbilt University's News Archives, 52
venting, xii, 34–35, 39, 43, 71, 114
vessel head, 7

Washington Evening Star, 50
waste heat, 11, 98
water, 46
water, contaminated, 38
water lancing, 38
water level, reactor, 24
Westinghouse, 76, 78, 109

white-tailed deer, 75
wind machines, 105
windmills, 105
wind power, 92, 104–5
Windscale, 16–17, 33
WKBO, 27, 49
    See also media coverage
wood, 92, 106
work, 91
work platform, 42, 43, 44

xenon, 31, 65
    xenon-133, 65

yeast, 44
yellowcake, 6
Yucca Mountain, 109

zeolite, 38
zircalloy, 8
zirconium, 25, 26